Insect Conservation - Challenges and Possibilities in a Changing World

Edited by Sigmund Hågvar and Frode Ødegaard

Published in London, United Kingdom

Insect Conservation - Challenges and Possibilities in a Changing World
http://dx.doi.org/10.5772/intechopen.111279
Edited by Sigmund Hågvar and Frode Ødegaard

Contributors
Amritpal Singh Kaleka, Bindu Gudi Ramakrishna, C.N. Thanu, Frode Ødegaard, Gopu Sushma, Jasti Sri Vishnu Murthy, Mani Chellappan, Navkiran Kaur, Ranjith M.T., Sigmund Hågvar

First published in London, United Kingdom, 2025 by IntechOpen
IntechOpen is the global imprint of INTECHOPEN LIMITED, registered in England and Wales, registration number: 11086078, 167-169 Great Portland Street, London, W1W 5PF, United Kingdom

For EU product safety concerns: IN TECH d.o.o., Prolaz Marije Krucifikse Kozulić 3, 51000 Rijeka, Croatia, info@intechopen.com or visit our website at intechopen.com.

British Library Cataloguing-in-Publication Data
A catalogue record for this book is available from the British Library

Insect Conservation - Challenges and Possibilities in a Changing World
Edited by Sigmund Hågvar and Frode Ødegaard
p. cm.
Print ISBN 978-0-85466-542-6
Online ISBN 978-0-85466-541-9
eBook (PDF) ISBN 978-0-85466-543-3

Meet the editors

Sigmund Hågvar is an emeritus professor in nature conservation at the Norwegian University of Life Sciences. Fifty years ago, when employed for some years in the Ministry of Environment, he argued for the need for insect conservation. He has worked with several insect groups and in quite different habitats, such as forests, alpine environments, and various types of soils. He has taught general nature conservation, landscape ecology, and nature management. He is still engaged in practical nature conservation work in collaboration with nature conservation organisations. He has edited several books on topics related to nature conservation.

Frode Ødegaard is an Associate Professor and curator of entomology at the NTNU-University Museum in Norway. He has extensive experience with entomological projects on biodiversity and taxonomy in Scandinavia, Central America, and Australia. He is a specialist in Coleoptera and Hymenoptera and has been working on biodiversity questions as well as revisions and descriptions of new species based on both morphological and molecular methods. He is the author of five scientific books about bumble bees, ants, wasps and dung beetles. His extensive experience includes involvement in several conservation working groups and a long record of successfully communicating scientific results to the public.

Contents

Preface

The time has come to include insects as an important element in nature conservation work. Without the diversity and abundance of insects, the world would have looked quite different. Every second species on this planet is an insect. They play important roles in many ecosystems, and man depends on their "ecosystem services" like pollination, decomposition and soil formation. Moreover, insects are important food for many birds and other organisms. Recently, alarming results have been published on a decline in insect abundance. Fragmentation and destruction of their habitats are the main reasons. In addition, climate change represents an increasing stress factor for the insect fauna. This implies a gradual increase in mean temperature and more frequent extreme weather situations. The combined effects of habitat loss and climate change are worrying and difficult to predict. Measures in insect conservation should be taken from a long-term perspective, using insight into landscape ecology and practising the precautionary principle.

Insect conservation is a demanding task. These animals have important ecological functions in a variety of habitat types: From coast to inland, from agricultural fields to forests and mountains, and from wetlands to dry landscapes. This book presents selected spotlights on the need for protection within quite different insect groups and habitats: Pollinators in nature and agriculture, wood-decomposing beetles in Nordic coniferous forests, wetland and pond insects, as well as soil-living species and the very specialized guests of ant nests. Major threats are described, and advice is given on practical conservation measures. The text is easy to read and understand for non-biologists, and several chapters are richly illustrated.

The purpose of this book is threefold:

Firstly, we want to inform and inspire scientists, students, people in general, organizations and politicians about the importance of conserving the diversity and abundance of insects.

Secondly, there is a need to prepare humans and nature, including the insect world, for the escalating climate problems and the biodiversity crisis.

Thirdly, we want to promote the serious message from the Montreal meeting in late 2022: Considerable areas should be set aside or restored for the sake of climate, biodiversity and ecosystem services. This aim has relevance for insect conservation.

We hope this illustrated book will fuel the growing insect awareness and initiate practical measures for their conservation. We are in an exciting phase where insects will hopefully soon be regarded as a self-evident element in nature conservation work. From the old slogan "save the whale", focusing on the world´s giants, it is time to focus on the tiny creatures. An updated slogan could be "save the insects that run the world."

Beaty and function in combination: One of the pollinators: The brimstone butterfly (Gonepteryx rhamni). Photo: Ove Bergersen.

Sigmund Hågvar
Faculty of Environmental Sciences and Natural Resource Management (MINA),
Norwegian University of Life Sciences,
Ås, Norway

Frode Ødegaard
Department of Natural History,
NTNU University Museum,
Trondheim, Norway

Section 1

Status and the Need for Awareness and Conservation Measures

Chapter 1

Introductory Chapter: Insects – The Small Creatures That Run the World

Sigmund Hågvar

1. Introduction

According to the International Union for the Conservation of Nature [1], more than 1 million insect species have been described, representing half of all known species on the planet. Insects can be extremely numerous, and they participate in many ecological processes. They are pollinators, decomposers, and food for birds, mammals, and fishes. Predators and parasites among them act as ecosystem stabilizers. For man, insects provide important ecosystem services. Their role as pollinators in human food production is crucial. Insects are also valuable in education and science. Furthermore, they represent a rich aesthetic source (**Figure 1**), and we love to use them in various types of artworks. Being a very old group, with a long evolutionary history, insects can be said to have a moral right to persist. Conservation of their diversity will allow for further evolution, while a mass extinction in our time will bring several evolutionary lines to an end. Considering their immense diversity and numbers, we could very well characterize our globe as "the insects' planet".

2. Two global crises

Two crises have become global: The biodiversity crisis with loss of habitats and species and the climate crisis with temperature rise and extreme climatic events. These crises are closely tied together and must be solved simultaneously. It does not help to stabilize climate if ecosystems collapse. Unfortunately, climate change is a direct threat to biodiversity. In December 2022, world leaders met in Montreal to discuss the biodiversity crisis. Their advice was ambitious: one-third of the terrestrial habitats, and a corresponding part of marine areas, should be given protection. In addition to preserving biodiversity, such large-scale conservation of habitats would retard climate change due to nature's ability to extract carbon dioxide from the atmosphere. In other words, the proposed efforts to protect nature would be a win-win situation. Moreover, one-third of the damaged nature should be restored. This is in accordance with United Nations' declaration that the period 2021–2030 should be the decade for restoration of nature's qualities and functions. Since there is generally a stronger political focus on climate problems than on biodiversity conservation, there is a great need to highlight the biodiversity crisis—including insects' crucial ecological role in our world.

 IntechOpen

Figure 1.
Aesthetic diversity: the beautiful small tortoiseshell (Aglais urticae) needs Urtica dioica plants for its larvae. This butterfly hibernates as adult and can be observed early in spring. Photo: S. Hågvar.

3. Insect conservation—a relatively new task

About 50 years ago, entomology was a topic for specially interested persons. Today, a consciousness is growing that these tiny animals belong to the organisms that keep ecosystems going—and make human life possible. Likewise, 50 years ago, few persons had heard about ecology—today, we know that ecological insight is a key to a sustainable future. While engineers have long been admired for their constructions, biologists are now called upon as advisors for constructing a sustainable future. To be a biologist has become a responsible task. Especially those whose expertise covers both entomology and ecology have a duty to contribute with advice—to people, to politicians, and to researchers.

In an international context, insect conservation has been taken seriously only for a few decades. Let us shortly look at the process of growing concern. In 1983, "The IUCN invertebrate red data book" was published [2]. Here, a few insects were included. Already, the next year, a book on insect conservation from an Australian perspective appeared [3]. After that, IUCN followed up with a beautiful book on threatened swallowtail butterflies of the world [4]. The Council of Europe produced a rather thorough "Charter on Invertebrates," highlighting the ecological role of small animals and the need for conservation measures [5]. That charter was presented as "A European Cultural Revolution." Three years later, the Council published an alert about vulnerable forest invertebrates depending on dead wood (**Figure 2**): "Saproxylic invertebrates and their conservation" [6]. Two large books followed in the 1990s: "The conservation of insects and their habitats" [7] and "Insect conservation biology" [8]. Between these two books, in 1992, United Nations (UN) arranged a Conference on Environment and Development in Rio de Janeiro. Here, two important international agreements were settled: The climate convention and the biodiversity convention. The latter obliged the signature countries to protect their

Figure 2.
Threatened forest diversity: the flat bugs (Heteroptera, Aradidae) are adapted to live in the narrow space under loose bark where they use their long proboscis to suck on fungi and sap in the decaying wood. This species, Aradus conspicuus, is bound to old trees of Fagus sylvatica. In Norway, the species is endangered. Photo: Karsten Sund.

biodiversity—including insects. This initiated the preparation of national red lists and political documents for how to take practical conservation measures.

Today, the amount of literature about insect conservation is rapidly increasing. An extensive book, treating many practical aspects, including landscape ecology, is "Insect Conservation. A global Synthesis" by Samways [9]. There are also dedicated journals, as "Journal of insect conservation," which started in 1997, and "Insect conservation and diversity" from 2008 onward.

4. A common future for man and insects: a hope with a warning

How can we shape a favorable future for both humans and insects? The answer is complex, ranging from global challenges and national obligations, to local care for ponds, hollow oak trees, or single species. There is, however, a superior prerequisite: Both climate change and habitat loss must be combatted simultaneously.

"Think globally and act locally" was the slogan from the international Rio meeting in 1992, when the two conventions on climate and biodiversity were formulated and agreed upon. About 30 years later, climate change is accelerating, and biodiversity is more threatened than ever before. Red lists on species and habitats are growing longer, both international lists and national ones. Red lists on species are useful, not only for making rescue plans for special species but also because they tell a lot about habitats that are under pressure.

A world without beautiful butterflies is not one we wish for our grandchildren. Still worse might be the lack of humming pollinators, which secure our food and maintain flower diversity. Extinction of species with unique adaptations would even be a great loss for science and for man's understanding of ecology and evolution. In our time, insect diversity and abundance are globally under double stress, from both habitat destruction and climate change. The necessary measures to counteract these forces are numerous and diverse, from improving water quality in a small brook to preserve extant forest landscapes that have resilience towards large-scale disturbances like wildfires or massive storm felling. Endemic species need to be rescued in their local environment, and fragmented populations should, if possible, be united through corridors and habitat restoration. It is always better—and usually cheaper—to protect real nature than trying to restore destroyed nature. Our thinking should be along two main lines: on ecosystem level and in a long-term perspective.

On a national level, there may be many alternative ways to protect and restore insect life. While certain countries may still have considerable areas left of intact nature, others are dominated by insect-hostile cities and monotonous agricultural fields. In either case, landscape ecology is a good guide to conservation and restoration. Rather detailed advice in landscape ecology for insect conservation was given in the books "Insect conservation biology" [8] and "Insect conservation, a global synthesis" [9]. A general advice for making an impoverished landscape more insect-friendly is to create a mosaic of habitats, connected with suitable corridors for dispersal. Furthermore, a varied and rich flora will be valuable, both for nectar availability and as food source for larval stages. Trees, bushes, and forests, as well as moist patches in the terrain can mute both long-term and sudden climate changes. Access to such a varied mosaic-formed landscape, preferably with a certain topography, would allow species to change their range. For instance, dragonflies on migration might need suitable ponds in more northern localities or at higher altitudes. New ponds can easily be constructed, and such new ponds are often rapidly colonized by various organisms. Wetlands can to a certain degree be restored, and the water quality improved. Weakly damaged bogs may be reimbursed by blocking channels and ditches and allowing the water level to rise. However, a bog developed during 10,000 years since the last glaciation cannot be recreated after being transformed to garden soil. Certain habitats like hollow oak trees takes hundreds of years to develop, so younger trees should be cared for and allowed to age undisturbed and in a sufficiently high density to support the unique hole-tree-fauna for long times. Remnants of forest stands may allow to increase in area and left for free development toward old trees and the production of dead wood elements (**Figure 3**).

Agricultural areas can to a certain degree be transformed back to an earlier habitat mosaic by opening brooks, planting trees, and establishing edge habitats rich in bushes and flowers. Pollinators might in this way recolonize a monotonous landscape and improve the yields. Even city environments can be made more insect-friendly through trees, flowers, and possibilities for wild bees to make nests. Here is a great potential for creativity and for positive contribution by people.

While the task of restoring nature has recently achieved much attention, nothing is more important than to identify and protect remnants of intact nature. To realize 30% protected land area will in many cases imply to conserve what is left of real nature, plus to "protect" areas that are more or less damaged or under restoration. It may imply that plans for the use of areas for buildings, roads, or other purposes must be revised. In practice, conservation measures for insects will overlap strongly with measures for conserving biodiversity in general.

Much has already been written about the importance of rescuing the globe's rich insect fauna. We know a lot about threats and what should be done, but our

Figure 3.
Also beetles are pollinators: a large and elegant longhorn beetle is the Musk beetle (Aromia moschata). It typically visits and pollinates composite flowers of the family Apiaceae (formerly Umbelliferae). Larvae tunnel within the wood of willow trees (Salix sp.) and need up to 3 years for their development. Photo: Ove Bergersen.

Figure 4.
Looking for butterflies. Photo: Morten Heldal Haugerud.

knowledge is still rather fragmentary about which measures are most important. The effects of climate change on insects are only partly understood. Research is still important, both on a general level and to tailor rescue plans for single species.

A necessary basis for improved insect conservation is a rise in general knowledge and awareness about insects. We must also ask: what are the obstacles that nature conservation meets, and how can we change "business as usual" in areal planning, areal consume, and areal fragmentation? Political support is needed, as well as economic contribution. Since humans depend upon ecosystem services like pollination, it should be possible to promote the conservation of the insect fauna. The aesthetic value of insects is often a door to concern and respect. Much human art reflects our appreciation for the insect beauty and their often-stunning appearances. An insect-rich, experience-rich, and beautiful world would be a precious gift for our descendants.

Conserving insect diversity has become an urgent task. But at the same time, it is utterly meaningful (**Figure 4**).

Author details

Sigmund Hågvar
Norwegian University of Life Science, Norway

*Address all correspondence to: sigmund.hagvar@nmbu.no

References

[1] Available from: IUCN.org. Gland, Switzerland

[2] Wells SM, Pyle RM, Collins NM. The IUCN Invertebrate Red Data Book. Gland, Switzerland: IUCN; 1983. 632 pp

[3] New TR. Insect Conservation. An Australian Perspective, Series Entomologica 32. Springer Dordrecht; 1984. 184 pp

[4] Collins NM, Morris MG. Threatened Swallowtail Butterflies of the World. Gland, Switzerland: IUCN; 1985. 401 pp. + 8 colour pages

[5] Pavan M. A European Cultural Revolution: The Council of Europe's Charter on Invertebrates. Strasbourg: Council of Europe; 1986. 51 pp

[6] Speight MCD. Saproxylic Invertebrates and their Conservation. Strasbourg: Council of Europe; 1989

[7] Collins NM, Thomas JA, editors. The Conservation of Insects and their Habitats. London: Academic Press; 1991. 450 pp

[8] Samways MJ. Insect Conservation Biology. London: Chapman and Hall; 1994. 358 pp

[9] Samways MJ. Insect Conservation: A Global Synthesis. Boston, MA: Cabi; 2020. 559 pp

Chapter 2

Insects in Anthropocene: Decline, Stress Factors, and Conservation Advice

Sigmund Hågvar and Frode Ødegaard

Abstract

We have entered Anthropocene, a period where man's activity has significantly influenced the globe's ecosystems. From several countries, a decline in the number of insects has been reported. The main reasons are habitat loss and fragmentation due to agriculture and urbanization, pollution by synthetic pesticides and fertilizers, biological factors like pathogens and introduced species, and climate change. Reduced insect numbers may affect pollination and other ecosystem services. Agricultural and forest pests may increase. Maintaining a diversity of habitats in the landscape is a simple but important thumb rule for preserving invertebrate diversity and viable populations in the long term. Deteriorated nature must be restored. A considerable increase in awareness about insect conservation is needed to save the majority of insects for the future.

Keywords: Anthropocene, climate change, conservation advice, decline, ecosystem service, habitat loss, insect conservation

1. Introduction

We have now entered a new epoch in the history of the globe: Anthropocene. It is a term that describes the major planetary transition in the mid-twentieth century caused by human impacts but the detailed definition is still debated [1]. Wikipedia defines it as a period where the activity of Man significantly influences the Earth's geology, landscape, limnology, ecosystems, and climate. One effect is a loss of biodiversity, including the high diversity of insects, a process that may threaten important ecosystem services.

2. Decline in insect populations

A review paper for the European Union by Goulson [2] had the following title: "Insect decline: an ecological armaggedon". From that paper, we cite: "There is broad agreement amongst scientists that insect declines are driven by a range of factors, including habitat destruction, climate crisis, light pollution, increasing fertilizer use, and the impacts of invasive species. Pesticides play a key role as well".

A growing number of studies report declining numbers of insects. From Germany, Hallmann et al. [3] reported a decline of 76% in flying insect biomass over 27 years, based on standard malaise traps in 63 nature protection areas. This general decline could not be explained by weather changes, land use practices, or habitat types. The authors expressed their worry about animals depending on insects as a food source, as well as about ecosystem functions on the landscape level. From the United Kingdom, the Netherlands, and Belgium, Warren et al. [4] reviewed long-running population data on butterflies. A decline in overall numbers was noted: About 50% reduction in the United Kingdom and the Netherlands, and about 30% in Belgium. In these three countries, between 8 and 29% of the resident butterfly species had gone extinct. Loss and degradation of habitats, together with chemical pollution, were considered the main causes of decline. The authors asked for policy changes to protect European butterflies, and insects in general. Climate change in addition to those factors forces many species to adjust their range. Some migrate north toward a colder climate. Migration is, however, risky, and susceptible species face new threats. Most often, ranges were simply constricted [5].

Biesmeijer et al. [6] reported parallel declines in pollinators and insect-pollinated plants in Britain and the Netherlands. The decline was based on data before and after 1980 and covered assemblages of bees and hoverflies (**Figure 1**). Habitat and flower specialists, univoltine species, and non-migrants showed the most frequent decline. Insect decline in Costa Rica was reported by Janzen and Hallwachs [7]. Field observations of caterpillars, their parasites, and their associates in the national park Área de Conservación Guanacaste have shown a gradual decline in species richness and density since the late 1970s, and strongest after about 2005. Climate change was assumed to be the main reason, with increasing ambient temperature and erratic rainfall or heated air masses. Wagner et al. [8] referred to alarming decreases in moth abundance and diversity across Europe and pointed to multiple stressors: both reductions in habitat quantity and quality due to land use changes, and climate change.

Figure 1.
Soaring elegancy: a hoverfly (Syrphidae), one of several insect groups that pollinate in nature, in agricultural fields, and in gardens. Photo: Jørn Bøhmer Olsen.

Based on a review of 73 historical reports of insect declines across the globe, Sánchez-Bayo and Wyckhuys [9] estimated that 40% of the world's insect species may go extinct during the next few decades. The underlying drivers, in order of importance, appeared to be: habitat loss due to agriculture and urbanization, pollution by synthetic pesticides and fertilizers, biological factors like pathogens and introduced species, and climate change. The authors called for a rethinking of agricultural practices and cleaning of polluted waters, both in agricultural and urban environments.

Outhwaite et al. [10] discussed how intensive agriculture combined with climate change could reshape insect biodiversity worldwide. Their main conclusion was that preserving natural habitats within agricultural landscapes will greatly benefit the long-term conservation of insect diversity.

Didham et al. [11] stressed that future monitoring of insect diversity and abundance should be more scientifically sound, with robust time series and the ability to account for potential artifacts, taking into account that insect population fluctuations are complex. They asked for technological advances in sampling and novel computational approaches.

3. Habitat loss and fragmentation as stress factors for insects

Living in a great variety of habitats, insects are worldwide subject to a large number of stress factors. Fragmentation, degradation, and loss of habitats are among the major problems (e.g. [12–14]). Man is continually consuming nature for roads, buildings, agricultural land, etc. (**Figure 2**). Conservation plans that safeguard the remaining parts of intact nature are important. Fortunately, several insects may survive within relatively small patches of nature, but small populations are vulnerable to stochastic events and genetic loss. Large populations are more resilient and adaptable. Today, urban sprawl consumes much nature over the whole world, as does the need for new agricultural areas. Historically, the large, continuous forests in Europe and North America have been cut and cleared, leaving only remnants up to our time.

Figure 2.
Fragmentation of natural forests by new roads, and the following activity on and along the roads, is a threat to biodiversity. Photo: Jørn Bøhmer Olsen.

Reforestation is an important challenge today. Many insect species depending on natural forest environments with old trees and a rich and continuous production of dead wood are probably already extinct, and many are on the brink of extinction today. Tropical forests, harboring the world's most unique and species-rich insect fauna, are fragmented and shrinking rapidly in area, due to a number of factors. This happens despite the international convention on biodiversity and warnings that tropical forests contribute to climate stability.

Pollinators are an insect group with an important ecosystem function and at the same time vulnerable to several threats. They often need a combination of flower diversity and abundance and at the same time various types of nesting places. Pollinators are sensible for agricultural chemicals and are attacked by various predators, parasites, and microorganisms. Modern production methods in forestry and agriculture leave a minimum of good insect habitats. To make agricultural areas more insect-friendly is an interesting challenge. Open, sunny and flower-rich habitats are often good insect habitats but also favorable for man to build houses. Wetlands, ponds, and brooks with special insect fauna are dried up or polluted. The introduction of alien species is also a stressfactor for insects in certain places.

Many insect species have larvae in the soil, especially among Diptera and Coleoptera. There are even wingless midges that live continuously within the soil. Moreover, soil is a substrate for pupating, resting, or hibernating in several insects. Soil is, however, a habitat heavily disturbed by man, especially in agricultural landscapes with tilling, use of chemicals, trampling, and compression by heavy machines. In a special section, within this book we look closer at soil as an insect habitat.

4. Pollution and various poisons as stress factors for insects

Literature on the effect of pollution and poisons on insects is large and rapidly increasing. Some examples of the topic are given here. Goulson [2] pointed to the fact that insecticides that shall protect plants from pests also harm beneficial insects like pollinators. Moreover, natural enemies of crop pests, like ladybirds or parasitoid wasps (**Figure 3**) that consume aphids, are also killed, so that aphid populations may increase to problematic levels. Even fungicides and herbicides can be harmful to insects.

Neonicotinoids deserve a special mention since they are among the most widely used insecticides in the world. They belong to the so-called systemic insecticides that are absorbed by all parts of the plant. According to Goulson [15], even pollen and nectar may contain enough to give reduced production in bumblebee colonies. Being soluble in water, neonicotinoids also leach into soil and waterways. They are therefore ecosystem contaminants. Siviter and Muth [16] concluded that neonicotinoids and other systemic insecticides are major contributors to beneficial insect declines. For instance, Meslin et al. [17] found that even low doses of the neonicotinoid clothianidin could reduce the ability in the male moth *Agrotis ipsilon* to respond to sex pheromones from the female.

Fipronil is another widely used insecticide. It is a broad-spectrum poison that disrupts the insect's central nervous system. It is used in field pest control for various insect groups in corn, horticultural crops, golf courses, commercial turf, as well as in flea control of pets [18]. Fipronil has been blamed for the spread of "colony collapse disorder" among bees, and it is even poisonous for several aquatic organisms.

A third example of an insecticide with potential detrimental side effects is sulfluramid. Being used as a formicide, it breaks down to perfluorooctane sulfonate (PFOS),

Figure 3.
Small helpers: a parasitoid wasp (Hymenoptera, Aphidiidae) laying eggs in living aphids. Such species can be used in biological control of aphids instead of insecticides. Photo: Norwegian University of Life Sciences.

which is very toxic, persistent, and worldwide restricted. It is one of several insecticides used in the urban environment and has entered drinking water in many places [19].

Air pollution may be harmful to insects. Raman [20] showed that pollutants in the air can be detrimental to insects. Anomalies were found in the bee's heart rate, which increased with pollution. Furthermore, the number of blood cells (hemocytes) in bees was reduced in rural areas. Duque and Steffan-Dewenter [21] considered air pollution to be a threat to insect pollination. They mentioned potential mismatches between flowering and pollinator activity and changes in pollinator attraction to flowers. Wang et al. [22] documented that short-term particulate matter could affect the olfactory perception in the antennae of insects. In urban environments, the antennae of houseflies (*Musca domestica*) were more densely covered with particulate matter than in non-urban environments.

Since insects are in close contact with toxic elements on and within plants, as well as in air, soil, and water, they have a great potential to function as bioindicators for environmental pollution [23, 24].

Even light can be regarded as environmental "pollution". In cities and suburbs, as well as in the countryside, many night-active insects are attracted to lamps and die or are unable to mate or find places for egg-laying. A local reduction in the moth population was documented by counting larval numbers [25]. They found that street lighting strongly reduced moth caterpillar abundance and affected caterpillar development as well. Unfortunately, modern LED lamps with a white light had a stronger negative effect on moth caterpillar abundance than conventional, yellow sodium lamps.

5. Climate change as stress factor for insects

Up till now, the physical and chemical activities of humans have been the main reasons for the decline in insect species and numbers. In our time, climate change has come as an additional threat. Effects are due to both a gradual increase in mean temperature and various extreme climatic events. Let us look closer at climate change as a growing stress factor for the insect fauna.

5.1 General about climate change as a threat to the insect fauna

Since climate change represents a direct threat to the insect fauna, climate must be stabilized. Conversely, since an intact nature helps to absorb carbon dioxide, nature protection contributes to climate stabilization. Therefore, the climate crisis and the nature crisis must be solved simultaneously.

In a review article titled "Scientists'warning on climate change on insects", Harvey et al. [26] pointed to possible long-term consequences of climate change for insects, ecosystems, and ecosystem services. As a basis for their analysis, they had access to forecast change in global climate parameters from now to the end of this century: temperature, growing season length, and precipitation, as well as assumed extreme ranges of climatic factors. The authors distinguished between the effects of a gradual rise in mean temperature on one hand and accompanying extreme weather events such as hot and cold spells, fires, droughts, and floods on the other hand. Detailed analyses accompanied by examples showed that gradual, long-term warming would affect insects in several ways: through range shifts, phenology, and species interactions. Extreme weather events could lead to heavy decline or extinction. Surviving species might achieve reduced fitness and fertility, reduced access to food, or be attacked by pathogens. These climatic effects come "on top" of anthropogenic stress factors, like degradation and fragmentation of habitats, urbanization, and chemical stresses via agriculture practices.

5.2 Species that change their distribution

A shift in distribution range may be a serious threat. New areas may be unsuitable, for instance, due to unacceptable habitat, lack of food, or the presence of predators. Furthermore, a human-dominated landscape may contain barriers in the form of agricultural fields, roads, urban areas, water, etc.

Hill et al. [27] analyzed the distribution of 51 British butterfly species in the light of twentieth-century climate warming. Based on old records, it could be documented that several species had moved to higher altitudes. Eleven southerly distributed species had expanded northward. Out of 35 species, 30 had failed to track recent climate changes due to lack of suitable habitat. It was estimated that the range for northern species would decline by 65%, and for southern species by 24%. For butterflies in the eastern Alps, Rödder et al. [28] found that climate change drove the species upward, toward the summits. This was the case especially for mobile and generalist species, in contrast to sedentary or specialist species. Mountain species seem to be vulnerable to temperature increases. The authors underlined the great value of historical surveys and museum collections as a reference to earlier distribution ranges.

Beetles show similar reactions to increased temperature (**Figure 4**). For instance, Brandmayr and Pizzolotto [29] found that over the past three decades, ground beetles in the Mediterranean mountains showed a strong uphill shift, especially among open land species. Local extinctions were observed around or above the tree line. The authors concluded that epigean and hypogean carabids responded faster than plants to temperature rise, and that these beetles were excellent indicators of climate change. In the Alps, many glaciers are melting rapidly, and cold-adapted invertebrates try to chase the retreating ice border [30]. Endemic species, for instance among carabid beetles and Collembola, have evolved to live around glaciers in this area. These unique species face local extinctions.

DOI: http://dx.doi.org/10.5772/intechopen.1007485

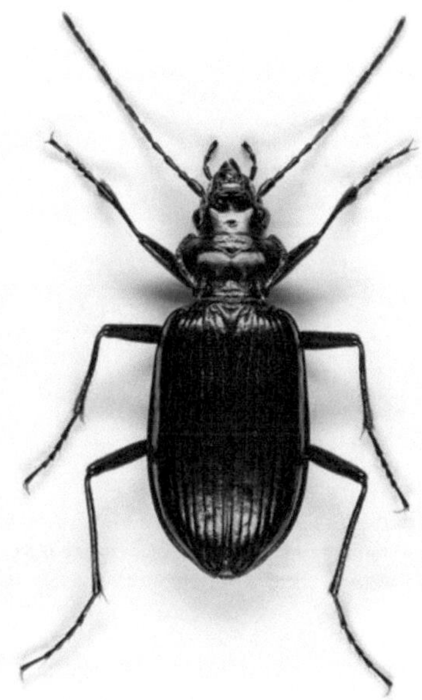

Figure 4.
*One of the specialists: cold-adapted insects may have problems due to climate change. This carabid beetle (*Nebria nivalis*) typically lives close to glaciers. If the glacier melts away, the beetle may disappear locally. Photo: Oddvar Hanssen.*

This exemplifies a typical road to extinction, where fragmented populations, one by one, disappear. Since this process takes time, we talk about "extinction debt" for species that may persist for some time but are "doomed" to gradually disappear.

Platts et al. [31] discussed 40 years of distribution data for nearly 300 British invertebrate taxa (aquatic bugs, bees, butterflies, dragonflies, damselflies, grasshoppers and allies, ground beetles, hoverflies, macromoths, non-marine mollusks, shieldbugs and allies, soldierflies and allies, spiders, and wasps). Dragonflies are good flyers and should be able to shift their distribution range easily. For example, in response to warming, the emperor dragonfly *Anax imperator* (**Figure 5**) has shifted its distribution northward and to higher elevations in Europe since 2000. Most variations in range shifts were, however, explained by a model that included both habitat and climate. In other words, a successful range shift will always depend on the availability of acceptable habitats. It means that management to help species to respond to future climates should focus on habitat protection and restoration. A general awareness of maintaining a diversity of habitats in the landscape is a simple but important thumb rule for preserving invertebrate diversity and viable populations in the long term.

5.3 Effects of climate change on phenology

During the two last decades of the twentieth century, the first appearance of most British butterflies was earlier than before this period [32]. The authors calculated that climate warming of the order of 1°C could advance the first appearance, as well as the peak appearance of most butterflies, by 2–10 days. Another publication 10 years later confirmed these findings [33]. Species with smaller range sizes had experienced

Figure 5.
*Changing distribution: the emperor dragonfly (*Anax imperator*) is the largest dragonfly in Europe. It has responded to a warmer climate by moving northward and to higher elevations. After 2013, the species has been recorded a few times in southernmost Norway. Illustration by Hallvard Elven.*

greater phenological advancement. Phenology changes may be risky for insects in several ways. One negative effect can be a mismatch between the activity period of the insect and the availability of nectar-producing flowers. Warmer winters have led to such a mismatch for honeybees [34]. While some insects may proliferate on higher temperatures and longer vegetation periods to achieve more generations yearly, such climate changes may represent a developmental trap for other insects, if they meet winter in a developmental stage that cannot survive or enter diapause [35, 36]. Warmer winters may affect different taxa in various, and often unpredictable, ways.

5.4 Effects of climate change on agricultural and forest insect pests

Increased temperatures and changing precipitation can affect insect pests in many ways. Their numbers may increase due to better survival during winter or because of the increased number of generations during the year. Their geographical distribution may expand, insect-transmitted plant diseases can be spread to new areas, and crop production could be at risk. In the future, we may need to monitor pest populations and try to develop new management tactics in agriculture [37].

Ma et al. [38] illustrated how a global agricultural insect pest, the diamondback moth *Plutella xylostella*, has been favored by climate change during the past 50 years and increased its overwintering range. Moreover, the pesticide resistance of the moth increased in overwintering sites, making the species an increasing threat to food production and the economy.

Similarly, new problems may arise in forest production. According to Pureswaran et al. [39], some recent outbreaks of bark beetles and defoliating insects have been due to climate change. Large outbreaks affect not only timber production but the whole forest ecosystem over considerable areas, including communities of forest insects.

5.5 Other effects, and the value of "insect refugia" in the landscape

Harvey et al. [26] referred to a number of other possible effects of climate change on insects. A temperature increase may, for instance, reduce their access to important food plants species, due to invasive plants or higher herbivorous pest populations. A changed climate may also lead to increased predation pressure or the establishment of new insect pathogens. Aquatic insects may be vulnerable to higher water temperatures due to lower oxygen content. The sensitivity to rising temperatures seems to be especially high in tropical aquatic insects [40].

In our considerations about the effects of climate change on insects, Harvey et al. [26] stress that we should distinguish between the effect of a gradual temperature increase and the occurrence of extreme events. Short-term and unpredictable climate extremes may be dramatic for the insect fauna, either it will be floods, drought, heat waves, or fires. While we, to a certain degree, may be able to forecast the effect of a slowly rising temperature on insects, unpredictable climate extremes, which might increase in frequency and impact, will make our predictions about future insect diversity and abundance difficult.

Ursul et al. [41] discussed closer how we could mitigate the effects of climate change on insects by establishing robust "refugia from climate change". So-called "macrorefugia" are heterogeneous landscapes, often in mountains, that permit local-scale redistributions and the evolution of endemics. "Microrefugia" provide high local environmental diversity that could buffer species against prevailing climate change by continuous change in their distribution pattern. For those who want to dive deeper into the topic, the chapter gives a rich literature list.

6. Restoration of nature qualities: Key messages from the European Union

An illustrating case study is that of the European Environment Agency [42]. Combined human pressures such as land degradation, resource overuse, and pollution reduce European nature quality on a broad base. There is a large need for ecosystem restoration. Such general measures would be favorable also for the insect fauna. From the referred publication, we cite the following key messages.

- In the EU, 81% of protected habitats, 39% of protected birds, and 63% of other protected species are in a poor or bad state. Only a very small fraction of these have shown any improvement over recent years.

- Diverse factors contribute to biodiversity loss, including land use, pollution, and climate change. Restoration efforts followed by ongoing management of the restored areas are needed both within and outside protected areas to ensure that our use of planetary resources is sustainable in the future.

- In the EU, 84% of crops at least partially depend on pollination by insects, and restoring pollinator habitats helps improve future food security.

- Improving and increasing the area of forests, wetlands, and seagrass meadows increases carbon sequestration and storage. Restoration improves ecosystem resilience, supporting nature-based production systems and helping them adapt to the increasingly frequent extreme weather events associated with climate change.

- Ecosystem restoration can improve health, well-being, and quality of life for people by increasing the availability of green spaces, mitigating pollution, and reducing the risk of diseases spilling over from animals to humans.

7. Practical advice in short

Much can be done to help insect diversity survive in a changing world. Insect conservation is a meaningful task, and its importance will increase in the coming years. Here is a concentrated list of practical advice.

1. *Identify and protect important insect localities.* Insects can have special habitat demands that are not covered by general conservation plans for various nature types. For instance, hollow oak trees harboring unique insect communities must be taken care of at the very place where they stand (**Figure 6**), specific ponds can be entomological hotspots, and certain specialized insects need ants as hosts. Certain butterflies or beetles may only thrive in a meadow landscape that is grazed by cattle. Communities of wild bees may need a combination of specific flowers and nesting places. Specialized cold-adapted insects live close to glaciers, while others prefer warm, south-faced hills. For the long-term conservation of the high diversity of forest insects, the ideal situation is to set aside large forest landscapes of many square-kilometer size and let nature work free. Saproxylic insects could then survive due to continuous production of dead-wood substrates. Insects depending on burned wood might, however, need artificial, controlled burning of forest patches.

2. *Practice the advice from the Montreal meeting in 2022 about the nature crisis.* This implies to decrease or halt anthropogenic stress on nature, including combatting threats from climate change. Set aside one-third of the land area for biodiversity and restore one-third of deteriorated nature.

Figure 6.
Insect conservation work ranges from landscape ecology to take care of single, valuable trees. This oak near Oslo has been protected for a long time due to its size and age. Its cavity also represents a habitat for many specialized insects. Photo: Anne Sverdrup-Thygeson.

3. *Restore deteriorated nature.* Various types of wetlands, from ponds and brooks to bogs and deltas, can sometimes be restored. Small forest patches can be allowed to increase in size and grow old. Formerly grassy fields colonized of bushes and trees can be cleared. A long-term restoration project could be to plant small oak trees and allow them to grow old and hollow. Even homogeneous agricultural landscapes may be restored to support better a diverse insect fauna (**Figure 7**).

Figure 7.
Historical loss of habitat diversity in a typical Norwegian agricultural landscape. Upper drawing: The situation around the year 1800, with a rich mosaic of habitats. Medium picture: The same landscape around the year 1930, with a reduced diversity of habitats. Lower picture: A typical situation around the year 1995, rather homogenized and with a low variation of habitats. Today, a restoration of some of the lost habitats, creating a better landscape mosaic, would greatly favor insect diversity. Illustrations: Vidar Asheim.

4. *Think landscape ecology.* To prepare insects (and other organisms) for an unpredictable future, general principles in landscape ecology should be practiced. In this respect, we recommend the book of Samways [43]. A landscape with a mosaic of habitats is often rich in species, including insects. Corridors between

Figure 8.
*Charismatic species can be door openers for engagement in insect conservation. This large beetle, the green rose chafer (*Cetonia aurata*) of the family Scarabaeidae, visits several flowers for nectar and pollen. The saprophagous, C-shaped larva has a 2-year life cycle and develops in leaf mold, manure, compost, or rotting wood. Adults are long-distance flyers. Photo: S. Hågvar.*

Figure 9.
We can all contribute: establishing a flower-rich garden supports insect conservation. Photo: Jørn Bøhmer Olsen.

Figure 10.
A new trend: a so-called "insect hotel", consisting of various types of straw and artificially bored holes in some kind of wood, can be attractive breeding habitats for various wild bees. Photo: Ove Bergersen.

patches will make dispersal easier. A mosaic landscape could buffer the harmful effects of climate change. Species might be able to adjust their living area according to increasing mean temperature and to survive extreme climate events. Improved knowledge about the ecology of red-listed species is valuable.

5. *Reduce chemical stress factors to insects*. Reduce the application of insect-hostile chemicals in gardens and in agriculture. Rely as far as possible on natural pest control. Even pollution of air may interfere with insect orientation. Protect water-living insects from pollution stress and communicate their value as indicators of water quality.

6. *Contribute to raising awareness about insect conservation.* This last point on the list is perhaps the most urgent one: Without a considerable increase in awareness about insect conservation, we will not be able to save the majority of insects for the future. The vital role of pollinators and the esthetic value of beautiful insects may hopefully trigger engagement among people, entomologists, organizations, and politicians (**Figure 8**). People can personally support insect diversity in many ways: by offering nectar-rich flowers in their garden (**Figure 9**), by constructing so-called insect hotels (**Figure 10**), by being aware of special insect habitats in their neighborhood (for example, ponds, old trees, or nesting sites of wild bees)—and by talking about the small creatures that run the world. Today, we are all responsible. If the leaders do not lead, it is the people (including entomologists) who must inform, inspire, and urge politicians to give priority to insect conservation. It does not help to stabilize climate if ecosystems collapse and important ecosystem services fail. A passive hope is not enough, a hope with a warning is needed, but the warning must be heard—and transformed into action.

Author details

Sigmund Hågvar[1*] and Frode Ødegaard[2]

1 Faculty of Environmental Sciences and Natural Resource Management, Norwegian University of Life Sciences, Ås, Norway

2 Department of Natural History, Norwegian University of Science and Technology, NTNU, Trondheim, Norway

*Address all correspondence to: sigmund.hagvar@nmbu.no

References

[1] Zalasiewicz J, Thomas JA, Waters CN, Turner S, Head MJ. What should Anthropocene mean? Nature. 2024;**632**:2024

[2] Goulson D. Insect Decline: An Ecological Armaggedon. Brusseles: Heinrich Böll Stiftung, European Union; 2022

[3] Hallmann CA, Sorg M, Jongejans E, Siepel H, Hofland N, Schwan H, et al. More than 75 percent decline over 27 years in total flying insect biomass in protected areas. PLoS One. 2017;**12**:e0185809

[4] Warren MS, Maes D, van Swaay CA, Goffart P, Van Dyck H, Bourn NA, et al. The decline of butterflies in Europe: Problems, significance, and possible solutions. Proceedings of the National Academy of Sciences. 2021;**118**:e2002551117

[5] Kerr JT et al. Climate change impacts on bumblebees converge across continents. Science. 2015;**349**:177-180. DOI: 10.1126/science.aaa7031

[6] Biesmeijer JC, Roberts SP, Reemer M, Ohlemuller R, Edwards M, Peeters T, et al. Parallel declines in pollinators and insect-pollinated plants in Britain and the Netherlands. Science. 2006;**313**:351-354

[7] Janzen DH, Hallwachs W. To us insectometers, it is clear that insect decline in our Costa Rican tropics is real, so let's be kind to the survivors. Proceedings of the National Academy of Sciences. 2021;**118**:e2002546117

[8] Wagner DL, Fox R, Salcido DM, Dyer LA. A window to the world of global insect declines: Moth biodiversity trends are complex and heterogeneous. Proceedings of the National Academy of Sciences. 2021;**118**:e2002549117

[9] Sanchez-Bayo F, Wyckhuys KA. Worldwide decline of the entomofauna: A review of its drivers. Biological Conservation. 2019;**232**:8-27

[10] Outhwaite CL, McCann P, Newbold T. Agriculture and climate change are reshaping insect biodiversity worldwide. Nature. 2022;**605**:97-102. DOI: 10.1038/s41586-022-04644-x

[11] Didham RK, Basset Y, Collins CM, Leather SR, Littlewood NA, Menz MH, et al. Interpreting insect declines: Seven challenges and a way forward. Insect Conservation and Diversity. 2020;**13**:103-114

[12] Haddad NM et al. Habitat fragmentation and its lasting impact on Earth's ecosystems. Science Advances. 2015;**1**(2). DOI: 10.1126/sciadv.1500052

[13] Hanski I, Pöyry J. Insect populations in fragmented habitats. In: Proceedings of the Royal Entomological Society's 23rd International Symposium, University of Sussex, UK, 12-14 September 2005. 2007. DOI: 10.1079/9781845932541.0175

[14] Kruess A, Tscharntkte T. Habitat fragmentation, species loss, and biological control. Science. 1994;**264**:1581-1584

[15] Goulson D. An overview of the environmental risks posed by neonicotinoid insecticides. Journal of Applied Ecology. 2013;**50**:977-987

[16] Siviter H, Muth F. Do novel insecticides pose a threat to beneficial insects? Proceedings B of The Royal

Society. 2020;**287**. DOI: 10.1098/rspb.2020.1265

[17] Meslin C et al. Sublethal exposure effects of the neonicotinoid clothianidin strongly modify the brain transcriptome and proteome in the male moth *Agrotis ipsilon*. Insects. 2021;**11**. DOI: 10.3390/insects12020152

[18] Gunasekara AS, Truong T, Goh KS, Spurlock F, Tjeerdema RS. Environmental fate and toxicology of fipronil. Journal of Pesticide Science. 2007;**32**(3):189-199. DOI: 10.1584/jpestics.R07-02

[19] Wee SY, Aris AZ. Revisiting the "forever chemicals", PFOA and PFOS exposure in drinking water. NPJ Clean Water. 2023;**6**:57. DOI: 10.1038/s41545-023-00274-6

[20] Raman S. Toxic pollutants in the air are taking the toll on pollinating insects. Research Matters. 11 Aug 2020. Available from: https://researchmatters.in/news/toxic-pollutants-air-are-taking-toll-pollinating-insects

[21] Duque L, Steffan-Dewenter I. Air pollution: A threat to insect pollination. Frontiers in Ecology and the Environment. 2024;**22**(3). DOI: 10.1002/fee.2701

[22] Wang Q et al. Short-term particulate matter contamination severely compromises insect antennal olfactory perception. Nature Communications. 2023;**14**:4112

[23] Chowdhury S et al. Insects as bioindicator: A hidden gem for environmental monitoring. Frontiers in Environmental Science. 2023;**11**. DOI: 10.3389/fenvs.2023.1146052

[24] Parikh G, Rawtani D, Khatri N. Insects as an indicator for environmental pollution. Environmental Claims Journal. 2021;**33**(2):161-181

[25] Boyes DH, Evans DM, Fox R, Parsons MS, Pocock MJO. Street lighting has detrimental impacts on local insect populations. Science Advances. 2021;**7**(35). DOI: 10.1126/sciadv.abi8322

[26] Harvey JA et al. Scientists'warning on climate change and insects. Ecological Monographs. 2022;**93**(1). DOI: 10.1002/ecm.1553

[27] Hill JK, Thomas CD, Fox R, Telfer MG, Willis SG, Asher J, et al. Responses of butterflies to twentieth century climate warming: Implications for future ranges. Proceedings of the Royal Society B. Biological sciences. 2002;**269**:2163-2171. DOI: 10.1098/rspb.2002.2134

[28] Rödder D, Schmitt T, Gros P, et al. Climate change drives mountain butterflies towards the summits. Scientific Reports. 2021;**11**:14382. DOI: 10.1038/s41598-021-93826-0

[29] Brandmayr P, Pizzolotto R. Climate change and its impact on epigean and hypogean carabid beetles. Periodicum Biologorum. 2016;**118**(3):147-162

[30] Gobbi M. Global warning: Challenges, threats and opportunities for carabid beetles (Coleoptera: Carabidae) in high altitude habitats. Acta Zoologica Academiae Scientiarum Hungaricae. 2020;**66**(Suppl):5-20. DOI: 10.17109/AZH.66.Suppl.5.2020

[31] Platts PJ, Mason SC, Palmer G, et al. Habitat availability explains variation in climate-driven range shifts across multiple taxonomic groups. Scientific Reports. 2019;**9**:15039. DOI: 10.1038/s41598-019-51582-2

[32] Roy DB, Sparks TH. Phenology of British butterflies and climate change. Global Change

Biology. 2001;**6**(4):407-416. DOI: 10.1046/j.1365-2486.2000.00322.x

[33] Diamond SE, Frame AM, Martin RA, Buckley LB. Species' traits predict phenological responses to climate change in butterflies. Ecology. 2011;**92**(5):1005-1012. DOI: 10.1890/10-1594.1

[34] Nürnberger F, Härtel S, Steffan-Dewenter I. Seasonal timing in honey bee colonies: Phenology shifts affect honey stores and varroa infestation levels. Oecologia. 2019;**189**:1121-1131

[35] Kerr NZ, Wepprich T, Grevstad FS, Dopman EB, Chew FS, Crone EE. Developmental trap or demographic bonanza? Opposing consequences of earlier phenology in a changing climate for a multivoltine butterfly. Global Change Biology. 2020;**26**:2014-2027

[36] van Dyck H, Bonte D, Puls R, Gotthard K, Maes D. The lost generation hypothesis: Could climate change drive ectotherms into a developmental trap? Oikos. 2015;**124**:54-61

[37] Skendžić S, Zovko M, Živković IP, Lešić V, Lemić D. The impact of climate change on agricultural insect pests. Insects. 2021;**12**:440. DOI: 10.3390/insects12050440

[38] Ma CS, Zhang W, Peng Y, et al. Climate warming promotes pesticide resistance through expanding overwintering range of a global pest. Nature Communications. 2021;**12**:5351. DOI: 10.1038/s41467-021-25505-7

[39] Pureswaran DS, Roques A, Battisti A. Forest insects and climate change. Current Forestry Reports. 2018;**4**:35-50. DOI: 10.1007/s40725-018-0075-6

[40] Tewksbury JJ, Huey RB, Deutsch CA. Putting the heat on tropical animals. Science. 2008;**320**:1296-1297

[41] Ursul G et al. Refugia from climate change, and their influence on the diversity and conservation of insects. In: González-Tokman D, Dáttilo W, editors. Effects of Climate Change on Insects. Oxford: Oxford University Press; 2024. DOI: 10.1093/oso/9780192864161.003.0016

[42] European Environment Agency. The importance of restoring nature in Europe. Briefing. 2023;**09**. DOI: 10.2800/269094

[43] Samways MJ. Insect Conservation. A Global Synthesis. Oxfordshire and Boston: Cabi; 2020, 559 pp

Chapter 3

Awareness for Insect Conservation, with a Short Presentation of Relevant Ideal Organizations

Navkiran Kaur and Amritpal Singh Kaleka

Abstract

Insects are essential to our planet. They are the small organisms that "run the world," at least in many ecosystems. Their ecosystem services are many: They pollinate plants and transfer seeds, and they are necessary for human food and agricultural growth and, as a result, have an effect on the economy. Every insect has a function to play in the global ecology, and the extinction of even a small number of species might have a large impact on biodiversity. Human removal of forests, expansion of agriculture, industrialization, and commercialization reduce insect diversity. Land use change, pesticides, and pollution seem to be the biggest threats to insects. However, climate change may eventually become a still larger threat. Here, we list various measures to raise the general awareness for insects and their needs for conservation. The quantity and quality of habitats are typically the main considerations in insect population support strategies. The creation, preservation, and reunification of fragmented ecosystems are additional strategies to raise sustainability. We list a number of organizations that work for protection of threatened insects and for insect diversity in general. These organizations are important also in raising the general awareness for, and respect for, the insect world.

Keywords: insects, conservation, species, diversity, habitat, awareness, organizations

1. Introduction

Insects play a significant role in the web of life. For humans, they offer various products such as food, honey, silk, dyes, and wax. They are often suitable in the study of basic biology and medicine. A large percentage of flowering plants rely on insects for fertilization and subsequent reproduction. Furthermore, insects play a crucial role in decomposition processes, and they serve as food for several vertebrates. The value of "ecosystem services" given by insects to humans and agriculture in the United States is estimated to be at least $70 billion in 2020 [1].

Even insects that we often consider pests can provide invaluable services. Termites, cockroaches, and blow flies, for example, help decompose dead and decaying organic materials, while mosquitos and caterpillars are vital food sources for many other organisms. Despite the fact that insects are acknowledged as vital ecosystem contributors, there is fear that their number and density may be declining globally as a result of destruction and loss of habitat, global warming, air pollution, and other factors [2]. There is plenty of justification to believe that these factors, along with growing human population and the process of urbanization, are causing losses in insect's diversity and numerous other species, even though evidence to support this is still sparse and contentious [3]. Better communication between insect conservation organizations and stakeholders, politicians, and managers of land is necessary.

Protecting, enhancing, or creating habitats across the landscape is essential to insect conservation because it allows even species with low dispersal ability to find a habitat when local conditions become unsuitable. However, insects can be affected by insecticides and other types of pollution, even if their natural habitats are preserved. In many cases, pesticides may be the only cost-effective way to manage the small percentage of species that are agricultural pests. Sadly, numerous non-target species—many of which are advantageous to farmers and growers—are harmed by this.

2. Strategies for awareness and conservation

The size and quality of habitats are main considerations in insect population management plans. Reunification of fragmented habitats can be important. Both agricultural stakeholders and the general population need to be aware of and understand the measures needed to support and enhance insect populations. Arousing public interest for insect diversity and their role in delivering ecosystem services is a worldwide task.

2.1 Preserving and creating insect-friendly habitats

As the world's natural habitats become more fragmented, insects will require high-quality habitats to be conserved and restored, along with pathways and travel corridors to facilitate their mobility. Forest habitats are always valuable, especially old forests with large trees and dead wood in different dimensions and phases of decomposition. Since many insects may live in small spaces, even a partial transformation of lawns to naturally occurring vegetation could contribute to the conservation of insects.

2.2 Minimize pesticide and herbicide use

Usage of pesticides destroys native insect populations, whereas lessening their usage encourages the growth of beneficial arthropods. A lot of pesticides are used purely for esthetic reasons, to make non-agricultural green areas like gardens, lawns, and parks look better. Both terrestrial and freshwater insect communities might benefit immensely from a decrease in or complete removal of cosmetic pesticide use [4]. Using strategies like integrated control, which combines several techniques to minimize or completely remove the dosage, can reduce the harm caused by pesticides. In situations where pesticides must be used, they need to be sprayed away from waterways, hedge rows, and other non-target areas.

2.3 Growing native plants

Native insects benefit from native plants more often than they do from nonnative ornamental varieties. For millions of years, native plants and native insects have coexisted in close ecological connections. Since they are adapted to the climate and weather conditions of their area, native plants may also require less maintenance. Insects can still benefit from growing a variety of nonnative plants, especially those that generate nectar, if native plants are not available. Native plants can be placed to patios, roofs, or areas between a city curb and sidewalk for residences without yards.

2.4 Limited use of lights

In the most biodiverse places on Earth, light pollution at night has doubled during the 1990s [5]. Lighting devices attract most nocturnal insects, and they are effective sensory traps that can cause fatigue in insects or lead to predation prior to daybreak [6]. Light pollution is probably to blame for the faster decline in nocturnal moths and butterflies in Europe compared to daytime flyers [7]. Given that fireflies utilize light to attract partners, LED lighting has also been demonstrated to decrease the success of their reproduction [8]. It has been shown that light pollution interferes with insect navigation, feeding, and courtship. Switch off your exterior lights at night, and urge the leaders of your local community to reduce the brightness of the road lighting when traffic is light. People should turn off lights when not in use, decrease light sources whenever needed, utilize motion-activated lighting, and switch to amber or red light bulbs, which emit frequencies that are less appealing to insects, in order to minimize the harm that they cause to insects [9].

2.5 Increasing awareness about insects

A major perceptual barrier to insect conservation is the general dislike and ignorance of insects, along with a lack of human recognition of their significance. Negative views on insects are common, and the general public in several countries is largely ignorant of the advantages and products that they offer. These impressions may be influenced by cultural beliefs that lack scientific support and can be enhanced by sensationalized media coverage, such as the use of dramatic headlines or movies that feature enormous, frightening insects. It takes a coordinated effort to change unfavorable stereotypes about insects. Making people aware of the advantages that insects offer to humanity is one approach to do it. Creating blogs and posting images of insects on social media are two strategies to raise appreciation.

3. Organizations working for insect conservation

Many organizations have insect conservation as one of their main goals. In the past, these organizations have typically been smaller than those that concentrate primarily on the more well-known animal species, particularly birds. Interest and involvement in this field are rising quickly as more people realize that insects and other invertebrates require much more attention. The main organizations that work specifically for insect conservation are as follows:

3.1 Buglife-the invertebrate conservation trust

In Europe, buglife is the sole organization dedicated to the preservation of all invertebrates. From bees to beetles, worms to woodlice, jumping spiders to jellyfish, they are actively trying to rescue our rarest tiny creatures. The headquarters of Buglife are located in Peterborough, England, and the company also maintains offices throughout the South West, Scotland, Wales, and Northern Ireland. Their mission is to keep invertebrate populations in the UK and abroad viable and to stop the extinction of invertebrates. They will preserve invertebrates and their natural environments, with a focus on conducting surveys to evaluate the condition of various species in particular habitat types, including freshwater, brownfields, and soft rock cliffs [10].

3.2 British bumblebee conservation trust

British Bumblebee Conservation Trust is an organization in the United Kingdom that works to protect and preserve bumblebees and their environment. It was founded as a result of significant concerns about the "plight of the bumblebee." The goal of the Trust is to establish a world in which bumblebees are treated with kindness and survive. Their goal is to spread and multiply bumblebee populations. An increasing number of devoted supporters are assisting their staff team in making a significant impact. The four primary goals of the Bumblebee Conservation Trust are as follows [11]:

- Expand knowledge about the ecology and conservation of bumblebees.
- Enhance the amount and quality of habitats for bumblebees.
- Motivate and empower a broad spectrum of individuals to take up bumblebee conservation.
- Be a long-lasting and successful organization.

3.3 Butterfly conservation

Butterfly Conservation

Butterfly conservation (BC) is a nonprofit environmentalist organization and charity operating in the United Kingdom with the mission of protecting the environment, moths, and butterflies. This nonprofit organization conducts programs to save more than 100 threatened Lepidoptera species and uses its research to offer guidance on how to preserve and restore habitats for butterflies and moths. In the UK, hundreds of butterfly and moth locations and reserves are being preserved by Butterfly Conservation [12].

3.4 Invertebrate link—A forum for relevant national organizations

Invertebrate link is the UK's forum for professional and volunteer organizations engaged in invertebrate conservation and research. The Committee currently has 36 member organizations represented on it, including governmental authorities, entomological groups, and top charities dedicated to conservation. By "facilitating information exchange between relevant organizations and statutory bodies, and by providing a context for cooperative ventures in relation to the development of strategy, policy, principles, and best practice," the Committee hopes to "advance the conservation of invertebrates in the UK [13]."

3.5 Amateur entomologists' society

Founded in 1935 by volunteers, the Amateur Entomologists' Society is a registered charity for people interested in insects (entomology). Additionally, a large number of individuals are generally interested in natural history. Their aim is to encourage entomology studies, particularly among young people and beginners. Additionally, the Society produces a variety of handbooks, brochures, and flyers on a wide range of subjects, such as Rearing and Studying Stick and Leaf-insects and Habitat Conservation for Insects [14].

3.6 British Dragonfly Society

British Dragonfly Society was established in 1983 with the intention of advancing research on dragonflies, damselflies, and their natural habitats, particularly in the United Kingdom. In order to gather Odonata records in the UK, the Society operates the Dragonfly Recording Network (DRN). Additionally, it supports many research and conservation initiatives [15].

3.7 British Entomological and Natural History Society

British Entomological and Natural History Society was initially referred to as the South London Entomological and Natural History Society and was established in 1872. The Society aims to promote and enhance entomology research, with a particular focus on the conservation of the United Kingdom's flora and fauna as well as the protection of species globally [16].

3.8 Royal Entomological Society

Originally called the Entomological Society of London, it was established in 1833. Insect research is the primary focus of the Royal Entomological Society. Its objectives are to increase communication among entomologists and to spread knowledge about insects [17].

3.9 Xerces Society for Invertebrate Conservation

Xerces Society for Invertebrate Conservation is named in honor of an extinct California butterfly, the Xerces blue (*Glaucopsyche xerces*) . The Society aims to promote invertebrate conservation, applied research, advocacy, public outreach, and education by collaborating with federal, state, and local organizations, including the US Department of Agriculture, as well as researchers, land managers, educators, and individuals. Through the preservation of invertebrates and their habitats, the worldwide charity Xerces Society for Invertebrate Conservation works to safeguard the natural world. Being a science-based organization, it carries out studies and uses the most recent data to direct its conservation efforts. Reducing pesticide use and its effects, protecting endangered species, and pollinator conservation are their main program areas [18].

3.10 Dipterists Forum

Dipterists Forum was founded in 1993. The society aspires to promote research into the documentation and preservation of true flies (Diptera). The objectives of the Dipterists Forum are as follows:

To promote the study of Diptera, establishing connections with other fields of research where there is a connection to other plants and animals.

To encourage the documentation of every facet of Diptera's natural history, including the development of distribution mapping.

To encourage Diptera conservation.

To promote and assist amateurs in museums, institutions, and universities while working in harmony with professionals.

To plan field trips, workshops, indoor meetings, and other pertinent events.

To distribute information *via* publications and newsletters.

To keep an interest in Diptera from continental Europe and beyond, but to concentrate on those from the British Isles [19].

3.11 The Bee, Wasp, and Ant Recording Society of the United Kingdom (BWARS)

Under the guidance of the UK Biological Records Centre (BRC), BWARS is a volunteer recording society that operates on a subscription basis. The Society is affiliated to the British Entomological and Natural History Society (BENHS). The Society aspires to establish connections with similar groups and interested parties across Europe in addition to encouraging the documentation of aculeate Hymenoptera in Great Britain and Ireland. The Society's objectives are to collect biological and distributional data on the aculeate Hymenoptera, which contains numerous significant pollinators, to educate and counsel members of the public and other society members and to further spread knowledge about aculeates [20].

4. Conclusion

Along with other invertebrates, insects are exceptionally successful and versatile in general. As a group, they may endure for as long as life on Earth can exist, but many species are extremely susceptible to abrupt changes brought about by the actions of humans. This renders many of the locations that would normally offer them adequate habitats unsuitable for their survival.

It is suggested that in order to increase public awareness, educators and environmentalists should emphasize the importance of insects to human survival and choose "icons" of insects, such as butterflies and bumblebees, to engage with the public. Primary forest sectors can operate as biodiversity reserves to replenish growing regions. In addition to a range of plants and trees, forests must also have fallen trees, leaf litter, logs, and branches, all of which provide microhabitats that are home to a diversity of insects.

Important reasons of insect population decrease are pesticides, herbicides, and other discharges from agricultural areas. Therefore, it is advised to switch to organic farming and agroforestry methods, which minimize pesticide use and keep it out of waterways. More broadly, insects may benefit from initiatives to combat climate change, preserve ecosystems, and decrease pollution.

Conflict of interest

The authors declare no conflict of interest.

Author details

Navkiran Kaur* and Amritpal Singh Kaleka
Punjabi University, Patiala, India

*Address all correspondence to: navkiran.dandiwal@gmail.com

References

[1] Losey JE, Vaughan M. The economic value of ecological services provided by insects. Bioscience. 2006;**56**:311-323

[2] Sánchez-Bayo F, Wyckhuys KAG. Worldwide decline of the entomofauna: A review of its drivers. Biological Conservation. 2019;**232**:8-27

[3] Crossley MS et al. No net insect abundance and diversity declines across US long term ecological research sites. Nature Ecology & Evolution. 2020;**4**:1368-1376

[4] Canadian Cancer Society. Cosmetic Pesticides: Information Brief. Alberta: Canadian Cancer Society; 2013. Available from: https://www.cancer.ca/~/media/cancer.ca/AB/get%20involved/take%20action/CosmeticPesticides-InformationBrief-AB.pdf

[5] Koen EL, Minnaar C, Roever CL, Boyles JG. Emerging threat of the 21st century lightscape to global biodiversity. Global Change Biology. 2018;**24**:2315-2324

[6] Frank KD. Effects of artificial night lighting on moths. In: Rich C, Longcore T, editors. Ecological Consequences of Artificial Night Lighting. Washington, D.C.: Island Press; 2006. pp. 305-344

[7] Coulthard E, Norrey J, Shortall C, Harris WE. Ecological traits predict population changes in moths. Biological Conservation. 2019;**233**:213-219

[8] Elgert C, Hopkins J, Kaitala A, Candolin U. Reproduction under light pollution: Maladaptive response to spatial variation in artificial light in a glow-worm. Proceedings of the Biological Sciences. 2020;**287**:20200806

[9] Frick TB, Tallamy DW. Density and diversity of nontarget insects killed by suburban electric insect traps. Entomological News. 1996;**107**:77-82

[10] Buglife—The Invertebrate Conservation Trust. Buglife | Help Buglife Save the Planet. Buglife. 2024. Available from: https://www.buglife.org.uk/

[11] Bumblebee Conservation Trust. The UK's Bumblebees Are in Crisis - Bumblebee Conservation Trust. 2024

[12] Home Page | Butterfly Conservation. n.d. Available from: https://butterfly-conservation.org/about-us

[13] Royal Entomological Society. Invertebrate Link—Royal Entomological Society. 2021. Available from: https://www.royensoc.co.uk/invertebrate-link/

[14] Amateur Entomologists' Society (AES)—The Gateway to Entomology. n.d. Available from: https://www.amentsoc.org/about/

[15] British Dragonfly Society—British Dragonfly Society. British Dragonfly Society. 2024. Available from: https://british-dragonflies.org.uk/

[16] Benhs. HOME—British Entomological & Natural History Society. British Entomological & Natural History Society. 2023. Available from: https://www.benhs.org.uk/about/

[17] Royal Entomological Society. Home—Royal Entomological Society. 2024. Available from: https://www.royensoc.co.uk/

[18] Xerces Society for Invertebrate Conservation. Home—Xerces Society for Invertebrate Conservation. 2024. Available from: https://www.xerces.org/

[19] Dipterists Forum—the Society for the Study of Flies (Diptera) | Dipterists Forum. n.d. Available from: https://dipterists.org.uk/about

[20] Watson A. Bees Wasps and Ant Recording Society. National Biodiversity Network; 2020. Available from: https://nbn.org.uk/biological-recording-scheme/bees-wasps-ant-recording-society/

Section 2

The Important Pollinators and Conservation Challenges

Chapter 4

Perspective Chapter: Wild Bees – Importance, Threats, and Conservation Challenges

Jasti Sri Vishnu Murthy, Bindu Gudi Ramakrishna, Mani Chellappan and Ranjith M.T.

Abstract

Wild bees hold tremendous significance as vital natural pollinators on a global scale. Approximately 20,000 bee species have been described worldwide. They are efficient pollinators owing to their species diversity and abundance, varied floral preferences, flight times, and reliance on weather conditions. Moreover, the extent and nature of pollination services provided by wild bees differ with geographical location, landscape type, climate conditions, and floral morphology. The decline of bees can be attributed to a combination of factors, such as loss, modification, and fragmentation of habitat, pesticide utilization, climate change, and the introduction of pests and diseases. Unlike honey bees, wild bees cannot easily be cultivated or reared in artificial conditions, hence strategies are needed to protect wild bees in the field. Conservation efforts can focus on protecting and restoring their natural habitats in different types of landscapes, implementing measures in human-altered environments, and utilizing human-made tools to support their well-being.

Keywords: wild bees, conservation, climate change, ecological resilience, pollination

1. Introduction

Insects are crucial pollinators; among them, bees, flies, beetles, and wasps are the significant pollinators of wild and cultivated plant species [1–3]. Bees are the foremost pollinators, with more than 20,000 species worldwide [4]. Moreover, honey bees are the main pollinators and are managed for crop pollination, though there are wild bees such as bumblebees and solitary bees (non-corbiculate and non-bombus wild bee species) being managed for improving pollination of specific crops [5–7]. Recently, wild bees have played a substantially larger role in crop pollination than honeybees. For instance, in the United Kingdom, wild bees account for 70% of crop pollination, with the remaining 30% being fulfilled by honeybees [8]. A similar study reveals that frequent visits by wild bees and hoverflies improve the fruit set of agricultural crops, even if honey bees visit regularly [9–11]. The crop yields are enhanced by the functional diversity of pollinators. Wild bees improve fruit and seed setting due to loose pollen held on body hairs and they often transfer more pollen than honey bees [12]. Wild bees are often important for wildflower pollination because rare flowers may rely on specific native wild bees for their existence. With decreased pollinator abundance, the rare flowers might not be adequately pollinated [13–17]. Hence, it is crucial to maintain our wild bee populations to ensure continued pollination of both crops and wildflowers [18]. However, over the century, insects have been declining at startling rates, including pollinators [19–22]. The number of wild bees has decreased everywhere [23], with 52 percent in the United Kingdom [14]. As much as 9.2% of wild bees are threatened according to the Red List in Europe [24]. Understanding the reasons behind these declines is crucial to mitigate the management measures to stop the loss of natural pollinators. In this chapter, we are focusing on opportunities and challenges in conservation of wild bees.

2. Why should wild bees have the utmost priority?

The significance of native, wild bees as pollinators for crops is a subject of ongoing debate, although their importance in natural ecosystems is generally recognized [25]. Approximately 80% of flowering plants rely on pollinators for fertilization, and wild bees play a predominant role as primary pollinators in the majority of ecosystems [1, 26–28]. The wild bees' forage in harsh climate conditions has high significance [29]. Bumble bees, in particular, forage at low temperatures, and certain flowers require high-frequency shaking, so-called sonication [30, 31]. This underscores the importance of wild bee pollination services, not only within the framework of ecosystem functioning but also in the context of global agricultural production. Many non-Apis bee species are as effective as or even better pollinators than honey bees for various crops [32–38]. The main challenge in utilizing these non-Apis species for crop pollination is not their effectiveness but rather their abundance [34–36]. Currently, management techniques are established for only a limited number of non-Apis bee taxa [39–41]. Furthermore, wild native bees that are not managed by humans also contribute to crop pollination as part of the ecosystem's natural services. In certain agricultural contexts, unmanaged bees can fully meet a crop's pollination requirements, and in other cases, they are frequent visitors to flowers, thereby aiding in fulfilling pollination needs for crops [42–44]. Additionally, when native bees coexist with honey bees, they can enhance the overall effectiveness of honey bee pollination [45–47]. Recognizing the role of native bees in pollinating crops serves as an important factor in garnering support for the conservation of bee populations.

3. Challenges in the conservation of wild bees

Wild bees play a critical role in maintaining the fragile equilibrium of ecosystems by pollinating a wide array of crop plants that sustain human existence. These bees are experiencing significant declines, and some may even be at the menace of extinction, with the status of many remaining uncertain and encountering a myriad of challenges that imperil their numbers and habitats. These challenges encompass multiple factors such as habitat loss, pesticide use, the impacts of climate change, and invasive species.

3.1 Habitat loss and fragmentation

Habitat loss stands out as the primary and widespread cause behind the global decline of bees. This loss of natural habitat is attributed to practices in agriculture, land abandonment, and urban development, resulting in fragmented and degraded habitats, which pose a significant threat to native bee populations. The removal of native vegetation primarily affects bees by reducing their access to both flowering plants and suitable nesting sites [48]. This reduction in the diversity and abundance of flowering plants puts considerable nutritional stress on bees, particularly wild ones that are more selective in their pollen preferences compared to honeybees [49–52]. Numerous bumblebee species, particularly those specialized in collecting pollen from leguminous plants, have suffered declines due to the loss of grasslands rich in flowers. Other semi-natural habitats like dunes, heathlands, and road verges have also deteriorated, impacting various bee species. Furthermore, the loss of large brownfield sites and the filling of quarries have exacerbated habitat loss. Habitat fragmentation isolates bee species, leading to reduced genetic diversity and increased vulnerability to diseases and parasites. Specialist species with limited dispersal abilities have experienced more substantial declines [14], resulting in the dominance of generalist species. Cleptoparasite species like the six-banded nomad bee and square-spotted mourning bee have seen dramatic declines, with some even facing extinction.

Changes in land use toward agriculture pose a significant threat to bee populations, resulting in the loss of grasslands and tropical forests. Intensive farming practices, such as the use of agrochemicals and soil plowing, can lead to soil degradation and the accumulation of harmful substances, further endangering the survival of both adult and larval bees [53, 54]. Reduced access to resources in agricultural landscapes negatively impacts the reproductive performance of solitary and bumblebees, leading to lower bee abundance and diversity. Ground-nesting stingless bee species like *Melipona quinquefasciata* are endangered due to activities such as firewood gathering and agricultural expansion, which destroy their nesting environments and floral host plants [55, 56]. The impact of urbanization on bumblebee populations is likely a complex and mixed one, with various studies suggesting both positive and negative effects [57–60].

3.2 Pesticide use and pollution

Insecticides are potentially the most harmful chemicals used in agriculture and play a significant role in the decline of wild bee populations [50]. However, the susceptibility of bee species to pesticides varies based on their behavior and natural characteristics. Different types of pesticides exhibit varying levels of toxicity to different bee species [61]. Pesticides can have detrimental effects on native bee populations in two ways: directly, by acting as insecticides that kill them, or indirectly, as herbicides that eliminate the plants they rely on for food. Sub-lethal effects of pesticides can be challenging to detect, but they have a more substantial impact on bee populations, primarily affecting the sensory perception, navigation abilities, and recognition of kin among wild bees.

Bees can be exposed to pesticides in multiple ways, as residues of insecticides are often found in the pollen and nectar of both cultivated crops and wildflowers [62, 63]. Bee species that have flight periods coinciding with pesticide applications and those relying on host plants vulnerable to pesticides face an elevated risk [64]. In managed grasslands, the use of herbicides reduces the variety and abundance of flowering plants, while in arable fields, herbicides that effectively control broad-leaved weeds can decrease the number of bees foraging for pollen and nectar. Many insecticides are absorbed by the lipids in pollen grains, potentially leading to the poisoning of solitary bee broods or causing mortality in the young bees of social colonies [65]. Systemic pesticides like neonicotinoids tend to have higher concentrations in pollen compared to nectar, potentially affecting different bee species depending on their foraging behaviors [66, 67].

Pesticide exposure has also been linked to various physiological problems in bees, including immunosuppression, reduced thoracic temperatures in *Osmia bicornis*, altered mitochondrial functions in *Bombus terrestris*, and a decline in the production of new queens in *Bombus terrestris* [68]. Furthermore, bees exposed to pesticides may experience reduced development of ovaries, decreased male fertility, lower production of offspring, a skewed male-dominated sex ratio, and increased mortality of eggs and larvae. These effects contribute to the decline of solitary bee populations in pesticide-intensive agricultural systems.

3.3 Climate change

Climate change represents an additional peril to wild bees, posing potential risks such as reduced numbers, shifts in habitats, and an increased menace of extinction, particularly for specialist or geographically limited species and small isolated populations. Climate change is also anticipated to lead the habitat loss, disrupt interspecies interactions, and jeopardize the survival and reproductive success of bees. Climate change may result in mismatches between the timing of flower blooms and bee emergence [69], leading to ecological imbalances where pollinators lose synchronization with their food plants due to shifting seasonal patterns. This could reduce the availability of floral resources for bees, particularly those with specific dietary preferences. In contrast, generalist species are expected to be less affected by these mismatches since they can visit multiple plant species.

Global warming led to a shift in the species such as the Yellow-legged mining bee, Hairy-footed flower bee, Buff-tailed bumblebee, Vestal cuckoo bee, Sharp-collared furrow bee, Painted nomad bee, and Dark blood bee to shift away from their traditional habitats [70]. Extreme weather events like droughts, floods, and storms may have a direct impact on bee populations. This is especially problematic for native bees as they cannot regulate their body temperature. Temperature fluctuations can affect various aspects of bee physiology and behavior. For instance, high temperatures during the summer can influence bee development, reproduction, and physical characteristics, including tongue length. Some bee species demonstrate phenotypic plasticity in response to climate change, leading to changes in species composition and traits. Warmer temperatures can also impact the size of bumblebee queens and the duration of winter hibernation, making them more vulnerable in the spring [57].

3.4 Invasive alien species

Invasive non-native species, diseases, and pathogens pose a significant risk to wild pollinators. Managed bees, like honeybees and bumblebees, can introduce harmful pathogens and parasites to wild bee populations, with the global increase in managed honeybee colonies and the importation of these bees being potential sources of exotic diseases for wild bees. Invasive alien species can, directly and indirectly, impact native biodiversity [71], raising concerns about the loss of pollination services for both wild plants and crops [14, 27, 72]. The consequences of invasive alien species can extend beyond pollination services, affecting entire ecosystems. Invasive alien plants often do not offer suitable rewards to native bees and can even be detrimental to them. Studying the impact of invasive alien plants on native bees at the individual level (changes in foraging behavior, survival, etc.) is relatively straightforward, but it's more challenging to assess their effects on bee populations and communities. Surprisingly, laboratory studies have shown that despite being separated for over 70 million years, honey bee pathogens can harm bumble bees [73–75], and vice versa [74]. These pathogens have also been found in various wild bee species, though their impact on species beyond Bombus remains poorly understood. Domesticated bumble bees and honey bees are the primary vehicles for the pathogens to spread to new areas, it's crucial to monitor and control these colonies to protect wild bee populations. Introducing new diseases is a major concern, and some declining wild bee species may suffer due to pathogen-related issues. For example, the rapid decline in Bombus sensu stricto is linked to infection by the potentially introduced fungal pathogen *Nosema bombi*, which is rarely found in co-occurring species from other Bombus subgenera [76]. The impact of invasive alien plants on native pollinators, especially bees, is not well-documented. Some invasive plants have flowers specialized for pollination by animals other than bees, making it difficult for native bees to access nectar and pollen. For instance, *Salvia splendens* has nectar concealed deep within long corolla tubes, which native bees struggle to reach [77]. In some cases, invasive alien flowers require specific behaviors that native bees lack. For example, native bees in Tasmania cannot handle the flowers of the invasive alien *Lupinus arboreus*, which necessitates large, powerful bees to expose stamens and stigma [78]. Similarly, native halictid bees cannot perform buzz pollination on the flowers of the invasive alien *Solanum torvum* because they cannot sonicate the anthers. Consequently, native bees tend to avoid these invasive alien plants as they aren't suitable food sources for them. We see that there are significant challenges that hinder the conservation efforts aimed at safeguarding wild bee populations. From habitat loss and pesticide use to climate change and disease, each challenge poses a unique threat, requiring careful consideration and proactive solutions. By understanding these challenges, we can begin to develop effective strategies to protect these crucial pollinators and the ecosystems that rely on them.

4. Opportunities for conservation of wild bees

4.1 Habitat protection and restoration

First and foremost, it is important to protect natural environments for wild bees and to recreate large natural habitats for a diverse ecosystem [79]. Recent research revealed that the decline of terrestrial insects is less in protected areas, which emphasizes the significance of habitat protection [79]. Hence, several methods are being used to describe the protected areas, some considering species diversity alone and species distribution along with changes in climate [18, 80]. Ecological Nich Models are employed to assess the protected areas in South America for bumble bees [81]. Ecological restoration of habitats within these protected areas can significantly increase the abundance and richness of wild bee populations across different landscapes and regions [82]. The hedgerows, grasslands, and woodland margins were possible attributes for enhancement, not only for solitary, ground-dwelling bees who forage a lot of blooming plants, but also for social, above-ground nesting bees that visit a few blooms in these environments [83]. Similarly in Brazil, rainforests may enhance bees' above-ground nesting, highlighting the significance of conservation efforts in a protected bee habitat [84]. The success of restoration efforts depends on understanding the habitat and resource preferences of the targeted wild bee species. Different habitat types can yield varying responses in wild bee abundance, depending on the ecological traits of the species [85]. Restoration methods, such as grazing and burning, can have both positive and negative effects on wild bee populations and are often context-dependent. These restoration efforts take place within the framework of LIFE plans supported by the European Union, such as LIFE Butterflies and LIFE in Quarries, which aim to restore habitats that indirectly benefit wild bees. While insect conservation is still in its early stages compared to bird and mammal conservation, some projects, like the Urban Bees LIFE project, focus on increasing bee abundance and diversity in urban environments [18]. However, such initiatives are exceptions rather than the rule. Wild bees are often absent from conservation programs at both the political and policy levels, primarily due to a lack of understanding of their requirements. There is a need to incorporate pollinators, especially in semi-natural habitats, into national and sub-national conservation programs, with a specific focus on identifying targeted species. As LIFE projects failed, similar programs such as Urban Bees LIFE are in action (www.urbanbees.eu) and are targeting management to increase the bee population and bio-diverse environment of local bees in potential urban and peri-urban areas [85]. Hall and Steiner [86] described that US state projects did not consider the significance of bees in comparison to vertebrates, leading to a lack of understanding of their needs and restoration actions.

4.2 Management of seminatural habitats

In agricultural ecosystems, semi-natural habitats offer vital supplies for wild bees. Floral resources in managed landscapes are dynamic and change across time and geography. Hence, higher levels of semi-natural habitats in agricultural landscapes are typically linked to higher pollinator abundance and richness [87] as well as improved pollination services to crops [88]. Various taxa of pollinators may be significantly influenced by the type, structure, and floral content of semi-natural habitats [89]. According to some recent research, flower-rich grasslands in Central European agricultural landscapes may have more diversified and plentiful wild bee communities than woody habitats like hedgerows and forest edges [89, 90]. Furthermore, compared to forest borders, seeded flower strips locally contribute more to the maintenance of populations of generalist wild bee species [91]. Furthermore, due to the varying flowering phenologies of the major plant species in these environments, the role of floral resources can change during the season [92, 93]. For instance, it has been demonstrated that certain bumblebee species monitor floral resources in various habitats during the course of the season [92]. As a result, they switch from woody plants, which primarily flower in spring, to herbaceous plants, which continue to bloom profusely in summer [94], which serve as their primary pollen source. According to Schlellhorn [95], conservation management should therefore take into account ways to encourage resource continuity across landscapes. Therefore, encouraging various semi-natural habitat types in landscapes may be essential to maintaining a variety of bee meta-communities throughout the season.

4.3 Management of urban and agricultural areas

Conservation efforts play a crucial role in urban areas, where more than 55% of the wild bee population resides. The process of urbanization tends to diminish the abundance and diversity of wild bees due to decreased food resources and nesting sites [96, 97]. When managing urban environments, it's essential to consider ecological, economic (cost), and logistical (implementation and sustainability) factors to make these often-neglected areas more bee-friendly [98]. Protecting areas near highly urbanized regions can act as buffers for wild bee populations. Initiatives are emerging that encourage the creation and management of bee-friendly habitats in agricultural settings. Roadsides, hedgerows, parks, and urban gardens are all vital habitats for wild pollinators, supporting a high diversity of species, including rare ones. Implementing bee-friendly plans through rooftop gardens, parks, and roadsides has led to increased native bee populations in Amsterdam [99]. This environmental interconnection is essential because harmful ecological fragmentation can negatively affect small bee species, especially in these areas. Ensuring habitat connectivity is crucial, as fragmentation can impede the dispersal of smaller bee species. Conservation strategies in agricultural areas can have a positive impact depending on the specific measures, target species, and landscape composition. Restoring floral resources, such as prairies, is an effective method for promoting wild bee diversity. Agro-environmental actions like flower strips have been embraced in Europe to increase biodiversity in intensively managed agricultural landscapes [100]. These actions have proven beneficial for bumblebees, honeybees, and hoverflies in Germany, Belgium, and England [93]. Increased flower supplies have notably improved the size, density, and population of bumblebees [93, 101]. The impact of AES (Agri-Environment Schemes) has been infrequently determined [102], and Geppert [103] observed the effects of organic practices and floral strips on bee population survival and development. Both of these actions were positively associated with pollinators' strength and the growth of bumblebee hives, but their efficiency depended on the surrounding landscape [64]. In England, Wood [93] also assessed Higher Level Stewardship farms (HLS) to experiment with the impact of cultivated flowers on native solitary bee populations. For example, honeybees and bumblebees were positively influenced by Phacelia sp., while solitary bees predominantly foraged on sunflowers and seed mixes of wildflowers [104, 105]. However, changes in floral resources among different bee environments can lead to stress, depletion of flower assets, and alterations in the crop-pollinator network [19, 106]. Agro-environmental actions like flower strips mainly benefit generalist species but may not adequately support diverse wild bee communities. The use of pesticides has varying effects on wild bees, and the sensitivity of solitary bees varies widely. It is often recommended to apply the precautionary principle and explore alternative pest management practices such as plant essential oils or biomolecules [18]. Integrated pest and pollinator management (IPPM) is suggested to integrate measures specifically benefiting pollinators into pest management. Sustainable strategies for agricultural landscapes are of utmost importance. Projects like EcoStack aim to enhance sustainability in European food production by considering ecological, economic, and social aspects. The Protecting Farmland Pollinators project in Ireland uses a scoring system to identify pollinator-friendly farming practices [85]. The Interreg-Sudoe Poll-Ole-GI project focuses on identifying effective methods, including green infrastructures, to support pollinator communities in Mediterranean crops like sunflowers and oilseed rape. These efforts in urban and agricultural areas are essential for conserving wild bees, promoting biodiversity, and ensuring the well-being of future generations. The nutritional quality and power of crops like *Brassica napus* can positively influence the abundance of bee populations [107, 108]. The variety of proteins and essential amino acids required for the growth and development of bee populations is crucial for bee species' health. This can be achieved through the use of floral resources available in their habitats, especially in environments where floral resources are scarce. Different agricultural technologies, such as friendly planting, can be tested to enhance the quantity and quality of resources, as seen with strawberry *Fragaria x ananassa* and borage *Borago officinalis* [109], although their effect on pollinator abundance has not been thoroughly studied. Describing the chemical toxicity's impact on honeybees and generalizing it to bumblebees and solitary bee taxa is challenging, as the sensitivity of solitary bees varies widely. The POSHBEE and European strategies can be beneficial in understanding how pesticides can affect native bee fauna and their synergistic effects with other factors contributing to their decline. However, it's important to note that pesticides can also impact bee pollinators and are not a specific response to safeguard the bee world [110, 111]. Egan [111] introduced a newly designed strategy called integrated pest and pollinators management (IPPM), which aims to integrate crop pollinators, and bio-control agents, and minimize pesticide use to enhance agriculture food production [112–118]. This strategy, which we propose to call "Pollinator and Integrated Pest Management Technology (PIPMT)," represents a holistic approach to managing pests below economic injury levels while promoting agriculture food production.

4.4 Nesting resources: bee hotels

Recent research has emphasized the need to enhance the availability of floral resources for pollinators but has often overlooked the equally vital aspect of providing nesting sites for these species [119]. Few studies have delved into the abiotic and biotic factors influencing nesting site selection and nesting success among different bee species [120, 121]. Endangered bees, including those with unique nesting behaviors like soil-nesting and cavity-nesting bees, such as carder bees and those that nest underground or in snail shells, can benefit from abundant nesting opportunities. It is recommended to provide ample nesting resources, including the establishment of Wild Bee Inns, especially for smaller bee species in various distribution areas (121). Maclvor and Packer [122] also introduced environmentally protective strategies to fulfill the nesting needs of wild bees. They emphasized that while approximately 50% of the occupants in bee hotels were newly introduced non-native bees, a concerning 75% of them were taken up by wasps. Worryingly, they observed a negative relationship between wild bees and the species found in bee hotels [123]. This research highlights the positive response of artificial nesting structures for native bees and underscores the importance of creating nesting sites to support the diversity of bee species crucial for crop pollination [119, 122]. It is essential to focus on the size of cavities in bee hotels, as different bee species have specific requirements for nesting [18, 85]. Smaller diameter holes can facilitate the nesting of many wild species, while larger holes may deter smaller native bees, as they are more likely to be occupied by larger non-native bees like M. sculturalis [123]. Whether it's small patches of bare soil installations, Wild Bee Inns, or bee hotels, the practical benefits of these nesting options should be subject to thorough investigation and analysis.

4.5 Combatting invasive alien species (IAS)

Invasive alien plant species can have varied effects on wild bees, with some experiencing positive, neutral, or negative impacts [18, 85, 124]. The outcome depends on factors such as ecological context and life history traits. The sensitivity of ecosystems to invasions is influenced by the degree of disturbance and resource availability [125–128]. Invasions often occur in habitats associated with human activities, where invasive plants can provide valuable food resources for generalist pollinators and restore ecological functions like pollination. However, they can harm specialist bee species with limited diet flexibility by competing with native flora and replacing plant species foraged by specialists. Efforts to combat invasive plants should be context-specific and guided by the precautionary principle, ranging from eradication to population control. A recent meta-analysis found that the impact of exotic plant species on pollinator abundance is case-specific, and there is no blanket rule that all exotic species are harmful or harmless. Invasive pollinators can directly compete for food and nesting resources, transmit pathogens, and indirectly affect food webs and plant communities [126]. For example, the Asian hornet Vespa velutina poses a potential threat to pollinators and escaped alien bumblebees used for agricultural pollination raise concerns about their impact on global diversity. The introduction of non-native bumblebee species can lead to mating with native species and the decline of native populations. Coordinated international measures to prevent biological invasions through importation policies and monitoring of invasive species are crucial. At the European level, a regulation to mitigate the effects of invasive alien species came into effect in 2015 [18, 85]. It defines preventive and curative measures for 66 invasive species to reduce their harmful effects on ecosystems. A classification system called the "Black List" categorizes invasive species based on their environmental impact. The IUCN Invasive Species Specialist Group (ISSG) works to reduce threats from invasive species through awareness, prevention, control, and eradication efforts at a global scale. Overall, managing invasive species is essential to protect native ecosystems and their associated species [18, 85].

5. Wild bees: an action for all

Effective communication and education are essential for establishing a solid basis for insect conservation. It is important to use clear terminology and unambiguous concepts in scientific communication. This includes specifying the taxonomic and geographic scale of research and clearly articulating the research's aim and results. To build support for insect conservation, several priorities in communication and education are highlighted:

1. Citizen science programs: Developing citizen science programs that combine education and training with data collection is crucial. Despite some biases in citizen science data, it allows for the monitoring of insect populations over long time periods and contributes to mapping a significant portion of species in an area. This approach has been particularly effective in monitoring pollinators, which have garnered growing public interest.

2. Education and training: Improving school, education, and training programs is vital to enhancing public knowledge and appreciation of insects. Educating the public about the importance of natural history observation and insect conservation can lead to positive behaviors and increased awareness.

3. Media ethics training: Scientists should receive training in media ethics to effectively communicate scientific methods and processes to the public. Clear and accurate communication is essential for engaging the public in nature conservation efforts.

4. Broad and accurate communication: Communication about insects and their conservation should be broad, reaching various audiences through scholarly literature, social media, and other channels. It should also emphasize accurate reporting of the geographic and taxonomic scales of research results.

5. Changing perceptions: Overcoming common misconceptions, such as the immediate association of bees with honey, hives, and stings, is a challenge in insect communication. The focus should shift toward linking bees with their roles as pollinators and wild species.

6. Resident engagement: Engaging urban residents in insect conservation efforts is essential for long-term success. Residents should feel safe and find conservation measures esthetically pleasing. Creating areas specifically dedicated to agroecosystems, such as "pocket prairies," can help meet resident preferences.

7. Practical support: Providing citizens and stakeholders with practical information, such as recommended plants to support bees based on their nutritional value and local context, can empower them to take effective conservation actions. Technical knowledge, like avoiding agrochemicals and following specific mowing/pruning schedules, is essential for integrating pollinators into green space management.

8. Public-private partnerships: Collaboration between NGOs, private businesses, and public authorities can lead to effective conservation actions. Examples include partnerships between a supermarket chain and an NGO in Austria, collaborations between fruit farmers and municipalities in Belgium, and partnerships involving beer brewers, NGOs, and public authorities.

In summary, effective communication, education, and collaboration among diverse stakeholders are crucial for building support and implementing successful insect conservation measures.

6. Conclusion

In conclusion, the urgency of species conservation necessitates a shift from only accumulating knowledge to taking concrete actions based on available evidence. Conservation experts must clearly define the target species for conservation efforts, recognizing the diverse ecological traits, and floral and nesting requirements of different species. While many conservation measures have been based on empirical evidence from honeybees and bumblebees, it is crucial to acknowledge that these findings may not apply universally to wild solitary bees, emphasizing the need for tailored approaches. One central aspect that demands attention is the redesign of floral resource mixes used in conservation programs. These mixes should be based on empirical studies identifying plant species that are foraged by the targeted bee species and matching their flying seasons and nutritional needs. Similarly, there is an urgent need to address nesting resources beyond bee hotels, which cater to only a fraction of bee species. Conservation programs should consider alternative floral resources, especially in anthropogenically-driven environments, but decisions must be grounded in strong empirical evidence. Furthermore, the way we manage bee-friendly areas, including finding alternatives to pesticides, needs to be rethought. The majority of habitat restoration studies have been conducted in North America and Europe, making it essential to test conservation measures in a wider range of habitats globally. Lastly, the communication and education about bees should be reshaped to move beyond associations with beehives and honey. The optimization of these actions and collaboration among conservation actors will enhance public awareness of biodiversity and ecosystem services, especially in urban areas where people are increasingly disconnected from nature.

Author details

Jasti Sri Vishnu Murthy*, Bindu Gudi Ramakrishna, Mani Chellappan and Ranjith M.T.
Department of Entomology, College of Agriculture, KAU, Vellanikkara, India

*Address all correspondence to: srivishnumurthy.j@gmail.com

References

[1] Ollerton J, Winfree R, Tarrant S. How many flowering plants are pollinated by animals? Oikos. 2011;**120**(3):321-326

[2] Linder HP. Morphology and the evolution of wind pollination. In: Owens SJ, Rudall PJ, editors. Reproductive Biology in Systematics, Conservation and Economic Botany. Proceedings of a Conference. English, Conference Paper, UK, Richmond. Kew, Richmond, UK: Royal Botanic Gardens (KRBG); 2-5 Sep 1996. pp. 123-135. 20001606453

[3] Bawa KS. Plant-pollinator interactions in tropical rain forests. Annual Review of Ecology and Systematics. 1990;**21**(1):399-422

[4] Michener CD. The Bees of the World. 2nd ed. Baltimore: The John Hopkins University press; 2007. 992 p. DOI: 10.56021/9780801885730

[5] Gruber B, Eckel K, Everaars J, Dormann CF. On managing the red mason bee (Osmia bicornis) in apple orchards. Apidologie. 2011;**42**:564-576

[6] Sheffield CS. Pollination, seed set and fruit quality in apple: Studies with Osmia lignaria (Hymenoptera: Megachilidae) in the Annapolis Valley, Nova Scotia, Canada. Journal of Pollination Ecology. 2014;**12**:120-128

[7] Zhang H, Huang J, Williams PH, Vaissière BE, Zhou Z, Gai Q, et al. Managed bumblebees outperform honeybees in increasing peach fruit set in China: Different limiting processes with different pollinators. PLoS One. 2015;**10**(3):e0121143

[8] Breeze TD, Bailey AP, Balcombe KG, Potts SG. Pollination services in the UK: How important are honeybees? Agriculture, Ecosystems & Environment. 2011;**142**(3-4):137-143

[9] Hoehn P, Tscharntke T, Tylianakis JM, Steffan-Dewenter I. Functional group diversity of bee pollinators increases crop yield. Proceedings of the Royal Society B: Biological Sciences. 2008;**275**(1648):2283-2291

[10] Garibaldi LA, Steffan-Dewenter I, Winfree R, Aizen MA, Bommarco R, Cunningham SA, et al. Wild pollinators enhance fruit set of crops regardless of honey bee abundance. Science. 2013;**339**(6127):1608-1611

[11] Campbell AJ, Wilby A, Sutton P, Wäckers FL. Do sown flower strips boost wild pollinator abundance and pollination services in a spring-flowering crop? A case study from UK cider apple orchards. Agriculture, Ecosystems & Environment. 2017;**239**:20-29

[12] Woodcock BA, Edwards M, Redhead J, Meek WR, Nuttall P, Falk S, et al. Crop flower visitation by honeybees, bumblebees and solitary bees: Behavioural differences and diversity responses to landscape. Agriculture, Ecosystems & Environment. 2013;**171**:1-8

[13] Forup ML, Memmott J. The restoration of plant–pollinator interactions in hay meadows. Restoration Ecology. 2005;**13**(2):265-274

[14] Biesmeijer JC, Roberts SP, Reemer M, Ohlemuller R, Edwards M, Peeters T, et al. Parallel declines in pollinators and insect-pollinated plants in Britain and the Netherlands. Science. 2006;**313**(5785):351-354

[15] Rollin O, Benelli G, Benvenuti S, Decourtye A, Wratten SD, Canale A, et al. Weed-insect pollinator networks as bio-indicators of ecological sustainability in agriculture. A review. Agronomy for Sustainable Development. 2016;**36**:1-22

[16] Gibson RH, Nelson IL, Hopkins GW, Hamlett BJ, Memmott J. Pollinator webs, plant communities and the conservation of rare plants: Arable weeds as a case study. Journal of Applied Ecology. 2006;**43**(2):246-257

[17] Jacobs JH, Clark SJ, Denholm I, Goulson D, Stoate C, Osborne JL. Pollination biology of fruit-bearing hedgerow plants and the role of flower-visiting insects in fruit-set. Annals of Botany. 2009;**104**(7):1397-1404

[18] Tanda AS. Wild bees and their conservation. Indian Journal of Entomology. 2022;**84**(3):726-736

[19] Nichols RN, Goulson D, Holland JM. The best wildflowers for wild bees. Journal of Insect Conservation. 2019;**23**:819-830

[20] Hallmann CA, Sorg M, Jongejans E, Siepel H, Hofland N, Schwan H, et al. More than 75 percent decline over 27 years in total flying insect biomass in protected areas. PLoS One. 2017;**12**(10):e0185809

[21] Ewald JA, Wheatley CJ, Aebischer NJ, Moreby SJ, Duffield SJ, Crick HQ, et al. Influences of extreme weather, climate and pesticide use on invertebrates in cereal fields over 42 years. Global Change Biology. 2015;**21**(11):3931-3950

[22] Ewald JA, Wheatley CJ, Aebischer NJ, Duffield S, Heaver D. Investigation of the Impact of Changes in Pesticide Use on Invertebrate Populations. York: Natural England; 2016

[23] Ollerton J, Erenler H, Edwards M, Crockett R. Extinctions of aculeate pollinators in Britain and the role of large-scale agricultural changes. Science. 2014;**346**(6215):1360-1362

[24] Nieto A. European Red List of bees, IUCN: International Union for Conservation of Nature. European Commission, IUCN European Union Representative Office, IUCN Species Survival Commission (SSC), IUCN Species Survival Commission (SSC), Bumblebee Specialist Group. Available from: https://policycommons.net/artifacts/1374072/european-red-list-of-bees/1988308/ [Retrieved: February 19, 2024]. CID: 20.500.12592/xhckcb

[25] Winfree R. The conservation and restoration of wild bees. Annals of the New York Academy of Sciences. 2010;**1195**(1):169-197

[26] Neff JL, Simpson BB. Bees, Pollination Systems and Plant Diversity. 1993. pp. 143-167

[27] Klein AM, Vaissière BE, Cane JH, Steffan-Dewenter I, Cunningham SA, Kremen C, et al. Importance of pollinators in changing landscapes for world crops. Proceedings of the Royal Society B: Biological Sciences. 2007;**274**(1608):303-313

[28] Free JB. Insect Pollination of Crops. 2nd ed. London: Academic Press; 1993. 684 p

[29] Brittain C, Kremen C, Klein AM. Biodiversity buffers pollination from changes in environmental conditions. Global Change Biology. 2013;**19**(2):540-547

[30] Heinrich B. “majoring” and “minoring” by foraging bumblebees, Bombus vagans: An experimental analysis. Ecology. 1979;**60**(2):245-255

[31] King MJ, Buchmann SL. Floral sonication by bees: Mesosomal vibration by Bombus and Xylocopa, but not Apis (Hymenoptera: Apidae), ejects pollen from poricidal anthers. Journal of the Kansas Entomological Society. 2003;**76**(2):295-305

[32] Javorek SK, Mackenzie KE, Vander Kloet SP. Comparative pollination effectiveness among bees (Hymenoptera: Apoidea) on lowbush blueberry (Ericaceae: Vaccinium angustifolium). Annals of the Entomological Society of America. 2002;**95**(3):345-351

[33] Richards KW. Comparative efficacy of bee species for pollination of legume seed crops. In: Matheson SLA, Buchmann C, O'Toole PW, Williams IH, editors. The Conservation of Bees. Vol. 18. London: Academic Press; 1996. pp. 81-103

[34] Parker FD, Batra SW, Tepedino VJ. New pollinators for our crops. Agricultural Zoology Review. 1987;**2**:279

[35] Heard TA. The role of stingless bees in crop pollination. Annual Review of Entomology. 1999;**44**(1):183-206

[36] Kevan PG, Clark EA, Thomas VG. Insect pollinators and sustainable agriculture. American Journal of Alternative Agriculture. 1990;**5**(1):13-22

[37] Winfree R, Williams NM, Dushoff J, Kremen C. Native bees provide insurance against ongoing honey bee losses. Ecology Letters. 2007;**10**(11):1105-1113

[38] Kremen C, Williams NM, Thorp RW. Crop pollination from native bees at risk from agricultural intensification. National Academy of Sciences of the United States of America. 2002;**99**(26):16812-16816

[39] Bohart GE. Management of wild bees. Beekeeping in the United States. 1967:411

[40] Torchio PF. Bees as crop pollinators and the role of solitary species in changing environments. VI International Symposium on Pollination. 1990;**288**:49-61

[41] Richards KW. Non-Apis bees as crop pollinators. Revue Suisse de Zoologie. 1993;**100**(4):807-822

[42] Klein AM. Steffan–Dewenter I, Tscharntke T. Fruit set of highland coffee increases with the diversity of pollinating bees. Proceedings of the Royal Society of London. Series B: Biological Sciences. 2003;**270**(1518):955-961

[43] Klein AM, Steffan-Dewenter I, Tscharntke T. Pollination of Coffea canephora in relation to local and regional agroforestry management. Journal of Applied Ecology. 2003;**40**(5):837-845

[44] Ricketts TH, Regetz J, Steffan-Dewenter I, Cunningham SA, Kremen C, Bogdanski A, et al. Landscape effects on crop pollination services: Are there general patterns? Ecology Letters. 2008;**11**(5):499-515

[45] Greenleaf SS, Kremen C. Wild bees enhance honey bees' pollination of hybrid sunflower. National Academy of Sciences of the United States of America. 2006;**103**(37):13890-13895

[46] Chagnon M, Gingras J, DeOliveira D. Complementary aspects of strawberry pollination by honey and IndigenQus bees (Hymenoptera). Journal of Economic Entomology. 1993;**86**(2):416-420

[47] Ollerton J, Johnson SD, Hingston AB, Waser N, Ollerton J. Geographical

Variation in Diversity and Specificity of Pollination Systems. Chicago: University of Chicago Press; 2006

[48] Kremen C, Williams NM, Bugg RL, Fay JP, Thorp RW. The area requirements of an ecosystem service: Crop pollination by native bee communities in California. Ecology Letters. 2004;**7**(11):1109-1119

[49] Galimberti A, De Mattia F, Bruni I, Scaccabarozzi D, Sandionigi A, Barbuto M, et al. A DNA barcoding approach to characterize pollen collected by honeybees. PLoS One. 2014;**9**(10):e109363

[50] Goulson D, Nicholls E, Botías C, Rotheray EL. Bee declines driven by combined stress from parasites, pesticides, and lack of flowers. Science. 2015;**347**(6229):1255957

[51] Vaudo AD, Tooker JF, Grozinger CM, Patch HM. Bee nutrition and floral resource restoration. Current Opinion in Insect Science. 2015;**10**:133-141

[52] Danner N, Keller A, Härtel S, Steffan-Dewenter I. Honey bee foraging ecology: Season but not landscape diversity shapes the amount and diversity of collected pollen. PLoS One. 2017;**12**(8):e0183716

[53] Ramalho JF, Amaral Sobrinho NM, Velloso AC. Contaminação da microbacia de Caetés com metais pesados pelo uso de agroquímicos. Pesquisa Agropecuária Brasileira. 2000;**35**:1289-1303

[54] Caldas ED, de Souza LC. Chronic dietary risk assessment for pesticide residues in Brazilian food. Revista de Saúde Pública. 2000;**34**:529-537

[55] Lima-Verde LW, Freitas BM. Occurrence and biogeographic aspects of Melipona quinquefasciata in NE Brazil (Hymenoptera, Apidae). Brazilian Journal of Biology. 2002;**62**:479-486

[56] Eardley C, Roth D, Clarke J, Buchmann S, Gemmill B. Pollinators and Pollination: A Resource Book for Policy and Practice. Agricultural Research Council (ARC); 2006. xv + 77 pp

[57] Williams NM, Ward KL, Pope N, Isaacs R, Wilson J, May EA, et al. Native wildflower plantings support wild bee abundance and diversity in agricultural landscapes across the United States. Ecological Applications. 2015;**25**(8):2119-2131

[58] Goulson D, Hughes W, Derwent L, Stout J. Colony growth of the bumblebee, Bombus terrestris, in improved and conventional agricultural and suburban habitats. Oecologia. 2002;**130**:267-273

[59] Chapman RE, Wang J, Bourke AF. Genetic analysis of spatial foraging patterns and resource sharing in bumble bee pollinators. Molecular Ecology. 2003;**12**(10):2801-2808

[60] Osborne JL, Martin AP, Shortall CR, Todd AD, Goulson D, Knight ME, et al. Quantifying and comparing bumblebee nest densities in gardens and countryside habitats. Journal of Applied Ecology. 2008;**45**(3):784-792

[61] Arena M, Sgolastra F. A meta-analysis comparing the sensitivity of bees to pesticides. Ecotoxicology. 2014;**23**:324-334

[62] Botías C, David A, Horwood J, Abdul-Sada A, Nicholls E, Hill E, et al. Neonicotinoid residues in wildflowers, a potential route of chronic exposure for bees. Environmental Science & Technology. 2015;**49**(21):12731-12740

[63] Rundlöf M, Andersson GK, Bommarco R, Fries I, Hederström V,

Herbertsson L, et al. Seed coating with a neonicotinoid insecticide negatively affects wild bees. Nature. 2015;**521**(7550):77-80

[64] Brittain C, Potts SG. The potential impacts of insecticides on the life-history traits of bees and the consequences for pollination. Basic and Applied Ecology. 2011;**12**(4):321-331

[65] Loper GM, Ross BH. Concentration of methyl parathion from Penncap-M in pollens of various lipid and oil contents. Environmental Entomology. 1982;**11**(4):925-927

[66] Dively GP, Kamel A. Insecticide residues in pollen and nectar of a cucurbit crop and their potential exposure to pollinators. Journal of Agricultural and Food Chemistry. 2012;**60**(18):4449-4456

[67] Goulson D. An overview of the environmental risks posed by neonicotinoid insecticides. Journal of Applied Ecology. 2013;**50**(4):977-987

[68] Brandt A, Hohnheiser B, Sgolastra F, Bosch J, Meixner MD, Büchler R. Immunosuppression response to the neonicotinoid insecticide thiacloprid in females and males of the red mason bee Osmia bicornis L. Scientific Reports. 2020;**10**(1):4670

[69] Memmott J, Craze PG, Waser NM, Price MV. Global warming and the disruption of plant–pollinator interactions. Ecology Letters. 2007;**10**(8):710-717

[70] IPBES. The assessment report of the intergovernmental science-policy platform on biodiversity and ecosystem services on pollinators, pollination and food production. In: Potts SG, Imperatriz-Fonseca VL, Ngo HT, editors. Secretariat of the Intergovernmental Science-Policy Platform on Biodiversity and Ecosystem Services. Bonn, Germany; 2016. 552 p

[71] Parker IM, Simberloff D, Lonsdale WM, Goodell K, Wonham M, Kareiva PM, et al. Impact: Toward a framework for understanding the ecological effects of invaders. Biological Invasions. 1999;**1**:3-19

[72] Ghazoul J. Pollen and seed dispersal among dispersed plants. Biological Reviews. 2005;**80**(3):413-443

[73] Liu H, Pemberton RW. Solitary invasive orchid bee outperforms co-occurring native bees to promote fruit set of an invasive solanum. Oecologia. 2009;**159**:515-525

[74] Graystock P, Yates K, Evison SE, Darvill B, Goulson D, Hughes WO. The T rojan hives: Pollinator pathogens, imported and distributed in bumblebee colonies. Journal of Applied Ecology. 2013;**50**(5):1207-1215

[75] Fürst MA, McMahon DP, Osborne JL, Paxton RJ, Brown MJ. Disease associations between honeybees and bumblebees as a threat to wild pollinators. Nature. 2014;**506**(7488):364-366

[76] Cameron SA, Lozier JD, Strange JP, Koch JB, Cordes N, Solter LF, et al. Patterns of widespread decline in north American bumble bees. National Academy of Sciences of the United States of America. 2011;**108**(2):662-667

[77] Corbet SA, Bee J, Dasmahapatra K, Gale S, Gorringe E, La Ferla B, et al. Native or exotic? Double or single? Evaluating plants for pollinator-friendly gardens. Annals of Botany. 2001;**87**(2):219-232

[78] Goulson D. Effects of introduced bees on native ecosystems. Annual

Review of Ecology, Evolution, and Systematics. 2003;**34**(1):1-26

[79] Van Klink R, Bowler DE, Gongalsky KB, Swengel AB, Gentile A, Chase JM. Meta-analysis reveals declines in terrestrial but increases in freshwater insect abundances. Science. 2020;**368**(6489):417-420

[80] Sobral-Souza T, Vancine MH, Ribeiro MC, Lima-Ribeiro M. Efficiency of protected areas in Amazon and Atlantic forest conservation: A spatial temporal view. Acta Oecologica. 2018;**87**:1-7

[81] Krechemer FS, Marchioro CA. Past, present and future distributions of bumblebees in South America: Identifying priority species and areas for conservation. Journal of Applied Ecology. 2020;**57**:1829-1839

[82] Tonietto RK, Larkin DJ. Habitat restoration benefits wild bees: A meta-analysis. Journal of Applied Ecology. 2018;**55**(2):582-590

[83] Carrié R, Andrieu E, Cunningham SA, Lentini PE, Loreau M, Ouin A. Relationships among ecological traits of wild bee communities along gradients of habitat amount and fragmentation. Ecography. 2017;**40**(1):85-97

[84] Ferreira PA, Boscolo D, Carvalheiro LG, Biesmeijer JC, Rocha PL, Viana BF. Responses of bees to habitat loss in fragmented landscapes of Brazilian Atlantic rainforest. Landscape Ecology. 2015;**30**:2067-2078

[85] Drossart M, Gérard M. Beyond the decline of wild bees: Optimizing conservation measures and bringing together the actors. Insects. 2020;**11**(9):649

[86] Hall DM, Steiner R. Insect pollinator conservation policy innovations at subnational levels: Lessons for lawmakers. Environmental Science & Policy. 2019;**93**:118-128

[87] Holzschuh A, Steffan-Dewenter I, Tscharntke T. How do landscape composition and configuration, organic farming and fallow strips affect the diversity of bees, wasps and their parasitoids? Journal of Animal Ecology. 2010;**79**(2):491-500

[88] Garibaldi LA, Steffan-Dewenter I, Kremen C, Morales JM, Bommarco R, Cunningham SA, et al. Stability of pollination services decreases with isolation from natural areas despite honey bee visits. Ecology Letters. 2011;**14**(10):1062-1072

[89] Bartual AM, Sutter L, Bocci G, Moonen AC, Cresswell J, Entling M, et al. The potential of different semi-natural habitats to sustain pollinators and natural enemies in European agricultural landscapes. Agriculture, Ecosystems & Environment. 2019;**1**(279):43-52

[90] Rivers-Moore J, Andrieu E, Vialatte A, Ouin A. Wooded semi-natural habitats complement permanent grasslands in supporting wild bee diversity in agricultural landscapes. Insects. 2020;**11**(11):812

[91] Ganser D, Albrecht M, Knop E. Wildflower strips enhance wild bee reproductive success. Journal of Applied Ecology. 2021;**58**(3):486-495

[92] Cole LJ, Brocklehurst S, Robertson D, Harrison W, McCracken DI. Exploring the interactions between resource availability and the utilisation of semi-natural habitats by insect pollinators in an intensive agricultural landscape. Agriculture, Ecosystems & Environment. 2017;**246**:157-167

[93] Eeraerts M, Van Den Berge S, Proesmans W, Verheyen K, Smagghe G, Meeus I. Fruit orchards and woody semi-natural habitat provide complementary resources for pollinators in agricultural landscapes. Landscape Ecology. 2021;**36**:1377-1390

[94] Bertrand C, Eckerter PW, Ammann L, Entling MH, Gobet E, Herzog F, et al. Seasonal shifts and complementary use of pollen sources by two bees, a lacewing and a ladybeetle species in European agricultural landscapes. Journal of Applied Ecology. 2019;**56**(11):2431-2442

[95] Schellhorn NA, Gagic V, Bommarco R. Time will tell: Resource continuity bolsters ecosystem services. Trends in Ecology & Evolution. 2015;**30**(9):524-530

[96] United Nations. United Nations, Department of Economic and Social Affairs. Available from: https://www.un.org/development/desa/publications/2018-revision-of-world-urbanization-prospects.html [Accessed: August 14, 2020]

[97] Kennedy CM, Lonsdorf E, Neel MC, Williams NM, Ricketts TH, Winfree R, et al. A global quantitative synthesis of local and landscape effects on wild bee pollinators in agroecosystems. Ecology Letters. 2013;**16**(5):584-599

[98] Kohler F, Verhulst J, Van Klink R, Kleijn D. At what spatial scale do high-quality habitats enhance the diversity of forbs and pollinators in intensively farmed landscapes? Journal of Applied Ecology. 2008;**45**(3):753-762

[99] Givetash L. Bees are dying at an alarming rate. Amsterdam may have the answer. NBC News. 2018:7

[100] Cole LJ, Kleijn D, Dicks LV, Stout JC, Potts SG, Albrecht M, et al. A critical analysis of the potential for EU common agricultural policy measures to support wild pollinators on farmland. Journal of Applied Ecology. 2020;**57**(4):681-694

[101] Wood TJ, Holland JM, Goulson D. Pollinator-friendly management does not increase the diversity of farmland bees and wasps. Biological Conservation. 2015;**187**:120-126

[102] Vaudo AD, Farrell LM, Patch HM, Grozinger CM, Tooker JF. Consistent pollen nutritional intake drives bumble bee (Bombus impatiens) colony growth and reproduction across different habitats. Ecology and Evolution. 2018;**8**(11):5765-5776

[103] Batáry P, Dicks LV, Kleijn D, Sutherland WJ. The role of Agri-environment schemes in conservation and environmental management. Conservation Biology. 2015;**29**:1006-1016

[104] Geppert C, Hass A, Földesi R, Donkó B, Akter A, Tscharntke T, et al. Agri-environment schemes enhance pollinator richness and abundance but bumblebee reproduction depends on field size. Journal of Applied Ecology. 2020;**57**(9):1818-1828

[105] Mallinger RE, Franco JG, Prischmann-Voldseth DA, Prasifka JR. Annual cover crops for managed and wild bees: Optimal plant mixtures depend on pollinator enhancement goals. Agriculture, Ecosystems & Environment. 2019;**1**(273):107-116

[106] Gérard M, Martinet B, Maebe K, Marshall L, Smagghe G, Vereecken NJ, et al. Shift in size of bumblebee queens over the last century. Global Change Biology. 2020;**26**(3):1185-1195

[107] Defra. The National Pollinator Strategy: For bees and other pollinators in

England. Department for Environment, Food and Rural Affairs. 2014

[108] Bukovinszky T, Rikken I, Evers S, Kleijn D. Effects of pollen species composition on the foraging behaviour and offspring performance of the mason bee Osmiabicornis (L.). Basic Applied Ecology. 2017;**18**:21-30

[109] Filipiak M. Key pollen host plants provide balanced diets for wild bee larvae: A lesson for planting flower strips and hedgerows. Journal of Applied Ecology. 2019;**56**(6):1410-1418

[110] Griffiths-Lee J, Nicholls E, Goulson D. Companion planting to attract pollinators increases the yield and quality of strawberry fruit in gardens and allotments. Ecological Entomology. 2020;**45**(5):1025-1034

[111] Dicks LV, Viana B, Bommarco R, Brosi B, Arizmendi MD, Cunningham SA, et al. Ten policies for pollinators. Science. 2016;**354**(6315):975-976

[112] Egan PA, Dicks LV, Hokkanen HM, Stenberg JA. Delivering integrated pest and pollinator management (IPPM). Trends in Plant Science. 2020;**25**(6):577-589

[113] Biddinger DJ, Rajotte EG. Integrated pest and pollinator management— Adding a new dimension to an accepted paradigm. Current Opinion in Insect Science. 2015;**1**(10):204-209

[114] Tanda AS. Entomophilous crops get better fruit quality and yield: An appraisal. Indian Journal of Entomology. 2019;**81**(2):227-234

[115] Tanda AS. Pollination efficacies of Apis mellifera L. and Tetragonula carbonaria (Smith) on peach. Indian Journal of Entomology. 2021;**83**(4):527-529

[116] Tanda AS. Insect pollinators matter in sustainable world food production. Indian Journal Entomology (Under publication). 2021

[117] Tanda AS. Biofloral phenology, foraging behaviour and entpollinatological effect of honey bees in pomegranate (Punicagranatum) fruit quality and yield. Journal of Horticultural Sciences. 2021;**8**(2):1-3

[118] Tanda AS. Urbanization and its impact on native pollinators. In: Advances in Insect Pollination Technology in Sustainable Agriculture. 23 Sep 2023

[119] Tanda AS. Native bees are important and need immediate conservation measures: A review. In: Proceedings of the 1st International Electronic Conference on Entomology. Basel, Switzerland: MDPI; 1-15 Jul 2021. DOI: 10.3390/IECE10523

[120] Fortel L, Henry M, Guilbaud L, Mouret H, Vaissière BE. Use of human-made nesting structures by wild bees in an urban environment. Journal of Insect Conservation. 2016;**20**(2):239-253

[121] Cane JH. A native ground-nesting bee (Nomia melanderi) sustainably managed to pollinate alfalfa across an intensively agricultural landscape. Apidologie. 2008;**39**(3):315-323

[122] Sardinas HS, Kremen C. Evaluating nesting microhabitat for ground-nesting bees using emergence traps. Basic and Applied Ecology. 2014;**15**(2):161-168

[123] MacIvor JS, Packer L. 'Bee hotels' as tools for native pollinator conservation: A premature verdict? PLoS One. 2015;**10**(3):e0122126

[124] Geslin B, Gachet S, Deschamps-Cottin M, Flacher F,

Ignace B, Knoploch C, et al. Bee hotels host a high abundance of exotic bees in an urban context. Acta Oecologica. 2020;**105**:103556

[125] Davis ES, Kelly R, Maggs CA, Stout JC. Contrasting impacts of highly invasive plant species on flower-visiting insect communities. Biodiversity and Conservation. 2018;**27**:2069-2085

[126] Burke MJ, Grime JP. An experimental study of plant community invasibility. Ecology. 1996;**77**(3):776-790

[127] Meerts P, Dassonville N, Vanderhoeven SO, Chapuis-Lardy L, Koutika LS, Jacquemart AL. Les plantes exotiques envahissantes et leurs impacts. Biodiversité: Etat, enjeux et Perspectives. 2004:238

[128] Alpert P, Bone E, Holzapfel C. Invasiveness, invasibility and the role of environmental stress in the spread of non-native plants. Perspectives in Plant Ecology, Evolution and Systematics. 2000;**3**(1):52-66

Section 3

Illustrating Spotlights on Selected Habitats and Insect Groups

Chapter 5

Three Quite Different Challenges in Insect Conservation: Spotlights on Odonata, Guests of Ants, and Soil Insects

Sigmund Hågvar

Abstract

Important insect localities may easily be overlooked in ordinary conservation plans. Odonata, ant guests, and soil insects illustrate three different approaches to their conservation. Odonata diversity can be limited by access to specific wetland or pond habitats, but their habitat demands can sometimes be restored. Ant guests depend fully on the long-term survival of their ant host species, which again depends on the preservation of sufficient habitat area. Soil insects may depend on a combination of specific soil types, vegetation, and climate for their larval development. Entomologists have a responsibility to identify critical ecological parameters for threatened insect species and to suggest tailored rescue plans.

Keywords: Odonata, ant guests, soil insects, habitat requirements, tailored conservation plans

1. Introduction

As explained in the introduction, this is a spotlight book with advice. We have already gone into some depth regarding the large functional group of dead wood-depending insects (the so-called saproxylic insects). In that section, we presented a case study of Nordic saproxylic beetles, their problems, and possible measures to conserve them. We have also treated pollinating insects, ground-dwelling carabid beetles, and the need for a general awareness for including insects in the general nature conservation work. Here we add three new spotlights, to further illustrate the diversity in insect conservation challenges: Odonata depending on a diversity of non-polluted freshwater habitats, the fascinating guests of ants, completely depending on the nests of their host species, and insects depending on suitable soil qualities, at least in part of their life cycle. Three ways of living, three types of vulnerability, and three examples on the overriding challenge of bringing insect diversity safe into the future.

2. The vulnerable Odonata (dragonflies and damselflies)

See (**Figure 1**).

The insect order Odonata consists of two groups: dragonflies (Anisoptera) and damselflies (Zygoptera). At rest, dragonflies keep their wings spread out or up, while damselflies keep their wings folded together along their body. As adults, both groups are predators, hunting other insects with great acrobatic maneuvering in the air. Even larvae are predators, but living in water, often in ponds or small brooks and rivers. On a world basis, Odonata comprises over six thousand species, with approximately equal species numbers in the two suborders dragonflies and damselflies (**Figure 2**).

2.1 Threats

Species are often sufficiently known to be evaluated as candidates for red listing. A press release from the international "IUCN Dragonfly Specialist Group" in December 2021 concluded that sixteen percent of the world's Odonata are red listed. More than half of these, eleven percent, belong to the three highest categories: critically threatened (CR), endangered (EN), or vulnerable (VU) [1]. IUCN concluded that:

"The destruction of wetlands is driving the decline of dragonflies worldwide, according to the first global assessment of these species. Their decline is symptomatic of the widespread loss of the marshes, swamps, and free-flowing rivers they breed in, mostly driven by the expansion of unsustainable agriculture and urbanization around the world".

The IUCN report pointed to water pollutants, eutrophication, and pesticides, as well as climate change, as current threats. However, positive changes have occurred. Due to improved water quality and restoring natural river morphology in Central Europe, many riverine species have recovered after heavy declines [2]. Such restoration success gives hope for the future.

Figure 1.
Escaped from water life: A newly hatched dragonfly clings to a straw above its empty pupa holster, slowly drying its wings. Photo: Jørn Bøhmer Olsen.

Figure 2.
Different colors, but one species: The so-called variable damselfly (Coenagrion pulchellum) in copulation. This European species is rare and declining in England but has stable populations for instance along the coast in Norway. The larva lives in vegetated ditches, canals, and ponds. Photo: Ove Bergersen.

2.2 The valuable but vulnerable ponds

While freshwater research, policy, and conservation up till now have focused mainly on larger water bodies, Hill et al. [3] stressed the need for taking care of small water bodies. Globally, ponds are among the most biodiverse and ecologically important freshwater habitats. The authors identified a number of knowledge gaps and research questions. Among them were anthropogenic stressors, pond monitoring, conservation management and policy, and the need for long-term monitoring of species communities and habitats. To halt the decline in freshwater biodiversity, we need to include pond ecosystems, and so-called pondscapes, in environment policy. A pondscape may be a pond surrounded by vegetation at a farm or in a garden, creating a beautiful and pleasant local world (**Figure 3**). On a landscape scale, connectivity is important, that is, ponds must be situated close enough to exchange species and individuals.

2.3 Odonata and climate change

A rising temperature in water bodies where Odonata larvae develop, or in the air where adults fly, can affect the developmental rate, phenology, and trophic interactions. Depending on dispersal ability and the availability of suitable habitats in the landscape, species may change their geographical distributions. A rapid

Figure 3.
Habitat diversity: This cultural landscape contains many different insect habitats: a pond, open fields, forests of different character and age, and belts of special vegetation between main habitats. The picture illustrates the potential of restoring habitats and landscape qualities. The pond has been artificially created, and the whole landscape may be managed to maintain habitat diversity. Photo: Ove Bergersen.

environmental change will initiate a strong selective pressure and a possible loss of populations and species. However, many species can benefit from the warming by expanding their range poleward. Physiological experiments may help to predict reactions to increasing temperature. The thermal sensitivity of Odonata may be used as a barometer for environmental change [4]. Following up on this idea, Cerini et al. [5] studied changes in the Odonata fauna during the last five decades in three countries (Tunisia, Mauritania, and Sweden). Whereas generalist species were often advantaged by warming due to their ability to colonize new habitats toward the north, specialists were more likely to go toward extinction. Species inhabiting lentic waters were more prone to show species turnover than species typical for standing waters. The authors called for more detailed long-term field studies of changes in local faunal composition.

Cadena et al. [6] predicted that dragonflies and damselflies in west and central Asia will undergo strong changes in diversity and distribution due to climate change. They estimated that the combined effect of anthropogenic forces and climate change will lead to near extinction of some species by 2100. In Europe, Cancellario et al. [7] forecasted widespread latitudinal and altitudinal rearrangements in Odonata community composition and traits due to climate change.

2.4 Conservation efforts and restauration possibilities

In an extensive study, Zhao et al. [8] studied the diversity and conservation needs of North American damselflies. Based on more than a hundred thousand georeferenced occurrence records of 296 damselfly species, species diversity, and environmental variables were mapped in a 100 x 100-km grid size. The most important parameters for species richness and endemism were water availability and temperature conditions. Five percent of the grids were selected as hotspots due to their species richness and weighted endemism. A considerable conservation gap was identified: Two-thirds of the hotspot grids were not a part of existing protected areas.

The diversity and conservation needs of European dragonflies and damselflies were studied by Kalkman et al. [9]. The central and western-central parts of Europe had the highest general species richness and of the predominantly lentic species. Concerning strictly lotic species, their diversity center was situated in southwest France and parts of the Iberian Peninsula. Endemic species were mainly found in the latter two areas as well as in the Balkan Peninsula. A strong mismatch was identified between species protected by the EU Habitats Directive, and today's conservation needs in the Mediterranean area. While water and habitat quality has improved since the 1990s in several central European countries due to early EU measures, the pressure on aquatic habitats has increased markedly around the Mediterranean. In Morocco, for instance, the marshes in the Smir area represent a rich Odonata locality but are today influenced by the tourism industry. A special study has been performed on the Odonata fauna in these wetlands and measures for protection have been considered [10].

Odonata may thrive in urban environments, even in artificial ponds, as documented in the capital of Oslo [11]. For instance, in the botanical garden, an ornamental pond has fostered enthusiasm for dragonflies and concern for their conservation.

The British Dragonfly Society [12] is very active, editing a Newsletter, mapping the distribution of species and arranging contacts, meetings, and excursions. They also run several conservation projects. One successful project has been to introduce the rare White-faced Darter (*Leucorrhinia dubia*) to Foulshaw Moss (**Figure 4**). The

Figure 4.
Extinction avoided: In England, the White-faced Darter (Leucorrhinia dubia) is a specialist of lowland bogs. However, most such habitats have been destroyed. By restoring bog habitats and translocating eggs and larvae to new localities, the species has now been rescued. Photo: Ove Bergersen.

Figure 5.
The vulnerable bogs: This bog in southern Norway, with an open pond in the central part, has offered a good living place for Odonata. Now the habitat is drained and destroyed, and the excavated turf is sold as a component of garden soil. Photo: Jørn Bøhmer Olsen.

Figure 6.
Successful restauration: This artificial pond in Norway has several functions. It absorbs organic pollution from agricultural fields, it is a rather good habitat for dragonflies, damselflies, and various plants, and it is a pleasant landscape element for humans. Photo: Ove Bergersen.

species is a specialist in lowland bogs, but peatland destruction (as in **Figure 5**) has limited the species to a few sites. In 2008, eggs and larvae were introduced into deep bog pools with floating Sphagnum moss in the Foulshaw Moss reserve. This translocation has been a great success. Drumburgh Moss is the next site into which the species will be introduced. Prior to the introduction, a number of bog pools have been created. Another endangered species in the United Kingdom is the Southern Damselfly (*Coenagrion mercuriale*), which has gone extinct, or is close to extinction, in seven European countries. Together with several other organizations, the British Dragonfly Society is working to save this beautiful blue species. A third project has been to conserve the rich dragonfly locality with several ponds in the Bramshill common, within Thames Basin Special Protection Areas, in North Hampshire. By modification, reshaping, and replanting the system of ponds, this work has dramatically improved the biodiversity of this pond landscape, both for dragonflies, other freshwater organisms, and rare plants. **Figure 6** illustrates a restored Norwegian pond with several functions.

3. Guests of ants: a vulnerable diversity in a modern world

3.1 Biology of ant guests

See (**Figures** 7 and 8).

Ants are social insects that may form large and durable communities. These colonies, or nests, are defended against predators, which may be other invertebrates, amphibians, reptiles, birds, or mammals. However, certain specialized animals, mainly among insects, have adapted to live within ant nests, taking favor of both the ants' defending ability and various habitats and food resources within the nest. These species have, during millions of years, evolved mechanisms to break through the ant'

Figure 7.
The nest of a host species: An ant mound of a Formica species, constructed mainly by spruce needles. Many ant guests live in such nests, which can be maintained for decades. Photo: S. Hågvar.

Figure 8.
The host: Busy Formica ants on the surface of an ant mound. Sometimes, especially in spring, some of the ant guests may appear on the surface of the nest. Others stay deep in the mound and must be excavated to be recorded. The ants show great ability to repair a damaged mound. Photo: S. Hågvar.

defense system. They may smell like ants, behave like ants, or excrete sugar or other chemicals that please the ants. Some even look like ants. Such "ant guests" represent many different insect groups, for instance, beetles (Coleoptera), flies (Diptera), butterflies (Lepidoptera), crickets (Orthoptera), spiders (Aranea), and other wasps (Hymenoptera). In Scandinavia, a total of 369 species of beetles are found in association with ants, of which 73 species may be characterized as myrmecophile [13]. Ødegaard et al. [14], list 123 species of myrmecophile arthropods found in Norway. Some of them live with little direct contact with the ants, for instance in the refuse heaps of the colony. Other guests are allowed to enter the innermost rooms, where they predate on ant eggs, larvae, or pupae. In certain cases, they are even fed by the ants. Among today's ant guests, we see various degrees of adaptations that may mirror the evolution process that the most sophisticated species have gone through (**Figure 9**).

The content of the present subchapter is mainly picked from a recent, fabulous book: *The guests of ants. How myrmecophiles interact with their hosts* [15]. The book goes deeply into recent research results, for instance how the ant world is to a high degree regulated by odors, and how certain guests produce chemicals that make them accepted by the ants. Among Staphylinidae beetles, for instance, there are species that not only smell like ants but may excrete chemicals that are very attractive to the ants, calming them down (**Figure 10**). If being attacked, certain beetles release chemicals that effectively repel the attacking ant.

A high level of adaptation to ant colonies is shown by larvae of certain species of blue butterflies of the family Lycaenidae. In early life, these larvae are plant eaters, but at a certain age, they drop to the ground. There, they are picked up by ants and carried into the nest. The trick used by the larva is to excrete a drop of sugar from a special gland on its back. The "valuable," sugar-producing caterpillar is carried to the brood chamber of the colony where it acts as a predator on ant larvae and eggs. The ants even feed them with regurgitated food, in the same way as they feed their own

Figure 9.
A very specialized beetle: The small and blind Claviger testaceus (Staphylinidae, Pselaphinae) is fully dependent on ants as a host. The beetle is allowed to feed on eggs and larvae in the brood chamber of Lasius or Myrmica ants. It pleases the ants by excreting sweet compounds from tuffs of hair [15]. Photo: Oddvar Hanssen.

larvae. In fact, the butterfly larva both smells like an ant larva and behaves like one. The relationship between the butterfly larva and the ant can be regarded as mutualistic since the larva provides sweet secretions for the ants. Pupation occurs within the ant nest, but the adult butterfly is not so welcome and leaves the colony rapidly.

3.2 Conservation aspects

In 1979, "the large blue butterfly," *Maculinea arion*, (also called *Phengaris arion*) went extinct in southern England. Attempts were made to reintroduce the species from mainland Europe, but without success. It was not until the detailed dependence on both a specific food plant, a specific ant species, and a specific microclimate that the reintroduction became a success. The food for the young larva was a *Thymus* plant, the host ant for the larger larva was *Myrmica sabuleti*, and the necessary management was to keep the vegetation short enough (for instance by grazing or periodic burning) to heat the soil sufficiently to make the actual ant species thrive. The direct cause of the decline of the butterfly during the 1970s was probably a disease (myxomatosis) that killed off the rabbits that had kept the grass short in the hillsides where *M. sabuleti* lived. In addition, the weather was unfavorable for the butterfly during some years, either too wet or too dry. A curious element in this story is that during the decline of the butterfly, fences had been set up to stop collectors. These fences toward lepidopterists also excluded grazing animals, worsening the ground climate for the ant. Finally, meticulous detective work over several years by scientists solved the complex riddle. This is now a classic rescue operation and has been described by Thomas et al. [16].

This case illustrates how difficult it may be to get a lost species back, but that detailed ecological insight can make it possible. Obviously, the best is to protect intact

Figure 10.
The staphylinid beetle Lomechusa emarginata depends on two different ants: It lives together with Formica ants during summer and with Myrmica ants during winter. From glands at the base of special abdominal hairs, the beetle secrets compounds that please the ants. The beetle is allowed to enter the ants' breeding rooms to deposit its eggs, and beetle larvae prey on eggs and larvae of the ant. Adults and larvae of the beetle are even fed by the ants. This relationship between ants and beetles is the result of a long evolution [15]. Photo: Oddvar Hanssen.

nature or to proceed with the usual management practice for species that depend on our cultural landscape.

While the referred book is a fascinating journey into the diversity and ecology of ant guests, it also illustrates their vulnerability in our modern world. These specialized guest species, being products of long evolutionary lines, fully depend on the continuous presence of their ant hosts. It is easy to foresee a massive loss of ant guests in our time, through loss of ant diversity by habitat destruction, or even climate change. There may, however, be situations where human habitat changes are favorable, for instance, for ant species that need open and sunny sites. The book opens and ends with a story told by Bert Hölldobler, about an abandoned limestone quarry in Germany, where he as a child learned a lot from his father about ants and their guests. Today, this species-rich quarry has been totally destroyed, being used as a landfill. The marvelous insect world that once inspired him has gone.

There is today a need to be conscious about the long-term conservation of ant species, either their habitats are open, sunny sites, wetlands and bogs, or various types of forest. Besides conserving intact and original nature, there are options to restore ant habitats, to create them artificially as in the mentioned limestone quarry which was an open, human-made habitat rich in surface stones under which several ant species thrived. Measures that preserve the diversity of ants, even conserve the highly specialized and fascinating guests. Without hosts, no guests.

From the book of Hölldobler and Kwapich [15], we learn that several guest species follow the migrating column of army ants, which are social hunters. Even certain specialized bird species take favor of the invertebrates scared up by the aggressive ants. Each colony of army ants needs a certain area of rainforest. Furthermore, the actual birds may follow different ant species during different periods of the year. For the long-term preservation of the ecological combination of ants, their guests, as well as the specialized bird fauna, a key question is: How large forest area is needed? A protected area must contain a minimum number of colonies of army ants as well as a certain number of army ant species. Skillful entomologists and ecologists are needed in such conservation work.

The future fate of insect diversity will to a high degree be a fight about area. This is especially evident when insect survival depends on the continuous presence of other species, either it may be a question of host plants or animal hosts. For ant guests, the host is not only a species of ant but also a whole community, or nest, with its complicated structure and function, microhabitats, and food sources.

We should add that even the book "The Ants" by Hölldobler and Wilson [17] has an interesting chapter about ant guests, listing the many invertebrates that have adapted to—and are dependent on—the continuous existence of ant colonies.

4. Soil insects: neglected diversity under our feet

See (**Figure 11**).

A "forgotten" topic in nature conservation is the rich fauna living in various soil types [18–23]. While insects that are active on the soil surface often attain attention, for example, the diversity of carabid beetles, the organisms under our feet are easily overlooked, although soil is teeming with life. Nowhere else in nature are species and specimens so densely packed as in soil. Here we find a mixture of insect larvae, springtails, mites, nematodes, enchytraeides, earthworms, and various other organisms. Among insects, especially Coleoptera and Diptera contain many species whose larvae are soil-living, either as root feeders, decomposers, or predators [19]. Sandy soils are also important nesting places for many species of Aculeate Hymenoptera, such as bees. Moreover, cracks in the soil or narrow spaces beneath stones may represent important hiding places, for instance for ground beetles during inactive periods. A related family to carabid beetles is tiger beetles (Cicindelidae) (**Figure 12**). The larvae live buried vertically in sandy soil, using their head to close the opening of their tunnel, but rapidly snapping a passing prey. Such open, sandy, and sunny habitats are vulnerable to human activity. Some tiger beetles are on the IUCN red list but most species are listed as data deficient.

Larvae of click beetles (Elateridae) are typically soil-living and often easy to identify due to their stiff body and yellow color. Several species of scarab beetles (Scarabaeidae) have their large, U-formed larvae in soil. For instance, larvae of the European cockchafer *Melolontha melolontha* (**Figure 13**) need three to four years for their development and feeding on roots. The scarab beetles belong also to a large number of dung beetles (often defined as a separate family, Geotrupidae). They are earth-boring beetles that transport animal excrements into soil, where the larvae develop (**Figure 14**). Furthermore, a large number of beetles from several families (Silphidae, Staphylinidae, Hydrophilidae, and others) contribute to bury or decompose dead birds or mammals lying on the soil surface. Most famous are the large burying beetles of the genus *Nicrophorus* within Silphidae. Beetles of the family Hydrophilidae also contribute in decomposing mammal excrements, a process that improves soil fertility (**Figure 14**).

Figure 11.
From soil to the air: Many Diptera species have soil-living larvae and pupae. This hatching crane fly is freeing its long legs from the pupa, which has just wiggled its way to the soil surface. Photo: Jørn Bøhmer Olsen.

Figure 12.
Tiger beetles (Cicindelidae) are famous for their running speed, their efficiency as predators, and their beauty. Larvae are sitting vertically buried in sandy soil. This is the common green tiger beetle (Cicindela campestris). Photo: Ove Bergersen.

Several groups of Diptera have soil-living larvae. The large, long-legged crane flies (Tipuloidea) are well-known among these. According to Frouz [24], soil-living Diptera larvae can be used as bioindicators, based on their ecological requirements and response to disturbance. Larvae of various Diptera groups occur in both natural and agricultural soil types. They can number several thousand per square meter and take part in many ecological processes. Furthermore, they respond to various stress

Figure 13.
A first flight: The forest cockchafer Melolontha hippocastani is soil-living as larva, feeding on plant roots and using 3–4 years for its development. Photo: Ove Bergersen.

Figure 14.
The Dung Beetle, or Dor Beetle (Geotrupes stercorarius), is a strong digger, transporting dung into the soil, as food for the larvae. Photo: S. Hågvar.

factors. Some specialized species among wingless Sciaridae and Cecidomyiidae dwell in soil during their entire life span.

Two other large insect groups are more or less soil-dwellers: ants (Formicoidea) and termites (Isoptera). Both groups may contribute significantly to the energy flow in the ecosystem, and to create a porous soil structure. An overview over termite biology, diversity, and sustainable management was given by Khan and Ahmad [25]. An opus magnus about ants is the book "The Ants" by Hölldobler and Wilson [17].

The many fascinating insects that are ant guests and depend on host ants for their existence, were described by Hölldobler and Kwapich [15]. Moreover, several insects need suitable soil for pupation or overwintering. For instance, adult larvae of many butterfly species dig down into soil for safe pupation, and/or overwintering.

Soil-dependent insects, and their habitat needs, are easily overlooked in insect conservation work. The soil fauna can have a rather local character depending on soil types, vegetation, and climate, and it is highly relevant to include different soil types as a parameter in biodiversity surveys. Intact soil profiles are vulnerable to changes in vegetation, trampling, or compression by heavy machines. Tilling changes the life conditions of soil animals dramatically. Primack [26] pointed to the "option value" of soil animals for practical use, for instance as decomposers, food, environmental indicators, or in science. Future insect diversity depends on our awareness of shielding a diversity of soils from human destruction, preserving their natural vegetation, and continuous litter production. In the long term, climate change represents a threat even to soil communities [19].

Author details

Sigmund Hågvar
Faculty of Environmental Sciences and Natural Resource Management, Norwegian University of Life Sciences, Ås, Norway

*Address all correspondence to: sigmund.hagvar@nmbu.no

References

[1] IUCN.org. Gland, Switzerland. Available from: https://www.iucnredlist.org

[2] Bowler DE et al. Winners and losers over 35 years of dragonfly and damselfly distributional change in Germany. Diversity and Distributions. 2021;**27**(8):1353-1366. DOI: 10.1111/ddi.13274

[3] Hill MJ et al. Pond ecology and conservation: Research priorities and knowledge gaps. Ecosphere. 2021;**12**(12):article e03853

[4] Hassall C, Thompson DJ. The effects of environmental warming on Odonata: A review. International Journal of Odonatology. 2008;**11**(2):131-153. DOI: 10.1080/13887890.2008.9748319

[5] Cerini F, Stellati L, Luiselli L, Vignoli L. Long-term shifts in the communities of odonata: Effect of chance or climate change? North-western Journal of Zoology. 2020;**16**(1):1-6

[6] Cadena JT, Boudot J-P, Kalkman VJ, Marshall L. Impacts of climate change on dragonflies and damselflies in west and Central Asia. Diversity and Distribution. 2023;**29**:912-925. DOI: 10.1111/ddi.13704

[7] Cancellario T, Miranda R, Baquero E, Fontaneto D, Martinez A, Mammola S. Climate change will redefine taxonomic, functional, and phylogenetic diversity of Odonata in space and time. npj Biodiversity. 2022;**1**(1). DOI: 10.1038/s44185-022-00001-3

[8] Zhao Z et al. Species diversity, hotspot congruence, and conservation of north American damselflies (Odonata: Zygoptera). Frontiers in Ecology and Evolution. 2023. DOI: 10.3389/fevo.2022.1087866

[9] Kalkman VJ, Boudot J-P, Bernard R, De Knijf G, Suhling F, Termaat T. Diversity and conservation of European dragonflies and damselflies (Odonata). Hydrobiologia. 2018;**811**:269-282

[10] Benazzouz B, Mouna M, Amezian M, Bensusan K, Perez C, Cortes J. Assessment and conservation of the dragonflies and damselflies (Insecta: Odonata) at the marshes of smir. Bulletin de l'Institut Scientifique, Rabat, section Sciences de la Vie. 2009;**31**(2):79-84

[11] Mauseth MI. Designs for Dragonflies – Odonata Diversity in Oslo, Norway [Master's Thesis]. Ås: Norwegian University of Life Sciences; 2018. 91 p

[12] England, Wales, Scotland: The British Dragonfly Society. Available from: british-dragonflies.org.uk

[13] Päivinen J, Ahlroth P, Kaitala V. Ant-associated beetles of Fennoscandia and Denmark. Entomologica Fennica. 2002;**13**:20-40

[14] Ødegaard F, Staverløkk A, Gjershaug JO. Maur i Norge. Kjennetegn, utbredelse og levesett. Trondheim: Norsk institutt for naturforskning; 2018. p. 447

[15] Hölldobler B, Kwapich CL. The Guests of Ants. Cambridge, Massachusetts and London: Belknap Harvard, The Belknap Press of Harvard University Press; 2022. 559 p

[16] Thomas JA, Simcox DJ, Clarke RT. 2009. Successful conservation of a threatened Maculinea butterfly. Science. 2009;**325**(5936):80-83. DOI: 10.1126/science.1175726. Epub 2009 Jun 18

[17] Hölldobler B, Wilson EO. The Ants. Berlin, Heidelberg, London, Paris,

Tokyo, Hong Kong: Springer-Verlag; 1990. 732 p

[18] Bennett A. The role of soil community biodiversity in insect biodiversity. Insect Conservation and Diversity. 2010;**3**:157-171

[19] Decaëns T, Jiménez JJ, Gioia C, Measey GJ, Lavelle P. The values of soil animals for conservation biology. European Journal of Soil Biology. 2006;**42**:23-38

[20] Decaëns T, Lavelle P, Jiménez JJ. Priorities for conservation of soil animals. CAB Reviews: Perspectives in Agriculture, Veterinary Science, Nutrition and Natural Resources. 2008;**014**:3

[21] Ertiban SM. Soil Fauna as webmasters, engineers and bioindicators in ecosystems: Implications for conservation ecology and sustainable agriculture. American Journal of Life Sciences. 2019;**7**(1):17-26. DOI: 10.11648/j.ajls.20190701.14

[22] Hågvar S. The relevance of the Rio-convention on biodiversity to conserving the biodiversity of soils. Applied Soil Ecology. 1998;**9**:1-7

[23] Robb C. The mainstreaming agenda of the convention on biological diversity and its value to protecting and enhancing soil ecosystem services. In: Ginzky H et al., editors. International Yearbook of Soil Law and Policy 2022. International Yearbook of Soil Law and Policy. Vol. 2022. Cham: Springer; 2024. DOI: 10.1007/978-3-031-40609-6_8

[24] Frouz J. Use of soil dwelling Diptera (Insecta, Diptera) as bioindicators: A review of ecological requirements and response to disturbance. Agriculture, Ecosystems and Environment. 1999;**74**(1-3):167-186. DOI: 10.1016/S0167-8809(99)00036-5

[25] Khan MA, Ahmad W. Termites: An Overview. In: Khan M, Ahmad W, editors. Termites and Sustainable Management. Sustainability in Plant and Crop Protection. Cham: Springer; 2018. DOI: 10.1007/978-3-319-72110-1_1

[26] Primack RB. A Primer of Conservation Biology. second ed. Sunderland: Sinauer Associates; 2000

Chapter 6

Perspective Chapter: Ground Dwelling Carabids – Challenges and Conservation in a Dynamic Environment

Bindu Gudi Ramakrishna, Jasti Sri Vishnu Murthy, Gopu Sushma and C.N. Thanu

Abstract

Ground beetles are a group of soil-dwelling insects belonging to the order Coleoptera and the family Carabidae. They exhibit great diversity in size and behavior with more than 40,000 species worldwide, of which 2000 are found in America and 2700 in Europe. Carabids have a wide range of feeding preferences, including generalists and specialized feeders. These beetles hold significant importance in conservation efforts due to their role as predators of insect pests and weed seeds, as bio-indicators, by contributing to ecosystem engineering and partaking in pollination. Occasionally some species act as pests of crop plants. Conservation of ground beetles faces several challenges, including habitat loss and fragmentation, pesticide use, agricultural practices, climate change, invasive species, and light pollution. Consequently, conservation strategies must be devised to safeguard ground beetles such as creating non-crop refuge habitats, employing targeted insecticides, adopting sustainable farming techniques, reducing light pollution, implementing monitoring programs, and engaging citizen science initiatives to help revitalize ground beetle populations in our ever-changing world.

Keywords: ground beetles, carabids, bioindicators, predators, conservation

1. Introduction

Ground beetles (Coleoptera: Carabidae) are a diverse and fascinating group of soil-dwelling insects with more than 40, 000 species worldwide [1]. They are exceptionally diverse and can be found in a wide range of terrestrial habitats, ranging from lush tropical rainforests to arid deserts. Their predominant nocturnal nature leads them to inhabit various niches, such as ground surface, soil, caves, or vegetation. Most carabid beetles exhibit omnivore tendencies, with a strong preference for predation [2], and play a pivotal role in the ecosystem by effectively controlling populations of various invertebrate pests including insects and other arthropods [3, 4]. This robust predatory nature positions them as valuable allies in natural pest management [5–9]. One of the most intriguing aspects of ground beetles lies in their remarkable ability to adapt and respond to various environmental fluctuations. Despite these ecological adaptations, ground beetles are facing numerous challenges such as habitat loss, pesticides, climate change, invasive aliens, etc. making them vulnerable and leading to the risk of extinction. In this chapter, we delve into the causes for the decline of ground beetles and discuss conservation strategies in a changing environment. Exploring the ground beetles unveils the intricate web of life on our planet and the interconnectedness of living organisms.

2. Diversity of ground beetles

The family Carabidae is one of the largest within the Coleopteran families, consisting of more than 40,000 documented species [3], 2142 genera, and 92 tribes found across the globe. India alone has 1900 known species, while Europe has more than 2700 species and North America boasts over 2000 species [1]. One of the most striking features of ground beetles is their remarkable diversity in size, shape, coloration, and behavior. Some are brilliantly colored, while others are cryptically patterned. The size of carabids ranges from a few millimeters to several centimeters in length. This diversity is not only visually captivating but also reflects their ability to occupy various ecological niches, from scavengers and predators to herbivores.

3. Habitat and feeding preference of carabids

The fundamental aspects that govern the relationship between organisms and their environments are habitat choice and feeding habits. Carabid beetles can be categorized into two groups based on their habitat preferences: widely distributed generalists, which can thrive in a variety of habitats, and specialists, which are limited to one or a few specific habitats [10]. Owing to their remarkable adaptability, carabids may be found in rather different habitats like forests, open fields, coastal areas, river banks, mountains, and peat bogs. Lebiini, predators of Chrysomelidae larvae, are particularly abundant in tropical forest canopies.

The feeding preferences of carabids are generally not fixed, and most of them are opportunistic feeders. They rarely feed on plant matter other than seeds. Some carabids are euryphagic, while others are stenophagic in nature. Based on their feeding habits, ground beetles can be categorized into granivores, carnivores, or omnivores (**Figure 1**).

Granivores: these species are of great importance, where both larvae and adult beetles feed on seeds (seed predators). Examples are the genera *Amara* and *Harpalus.*

Omnivores: these species are known to feed upon insects like aphids, caterpillars, grubs, and weed seeds. Examples are the genera *Pterostichus* and *Poecilus.*

Carnivores: These beetles are specialized predators, preying on a variety of small invertebrates, including insects, spiders and other arthropods. They have specialized mouthparts with sharp mandibles to capture and consume prey. Examples are *Calosoma, Carabus, Scaphinotus and Spaeroderus.*

Habitat preference and feeding behavior provide valuable insight for conservation actions to support ground beetles and their associated epigaeic fauna.

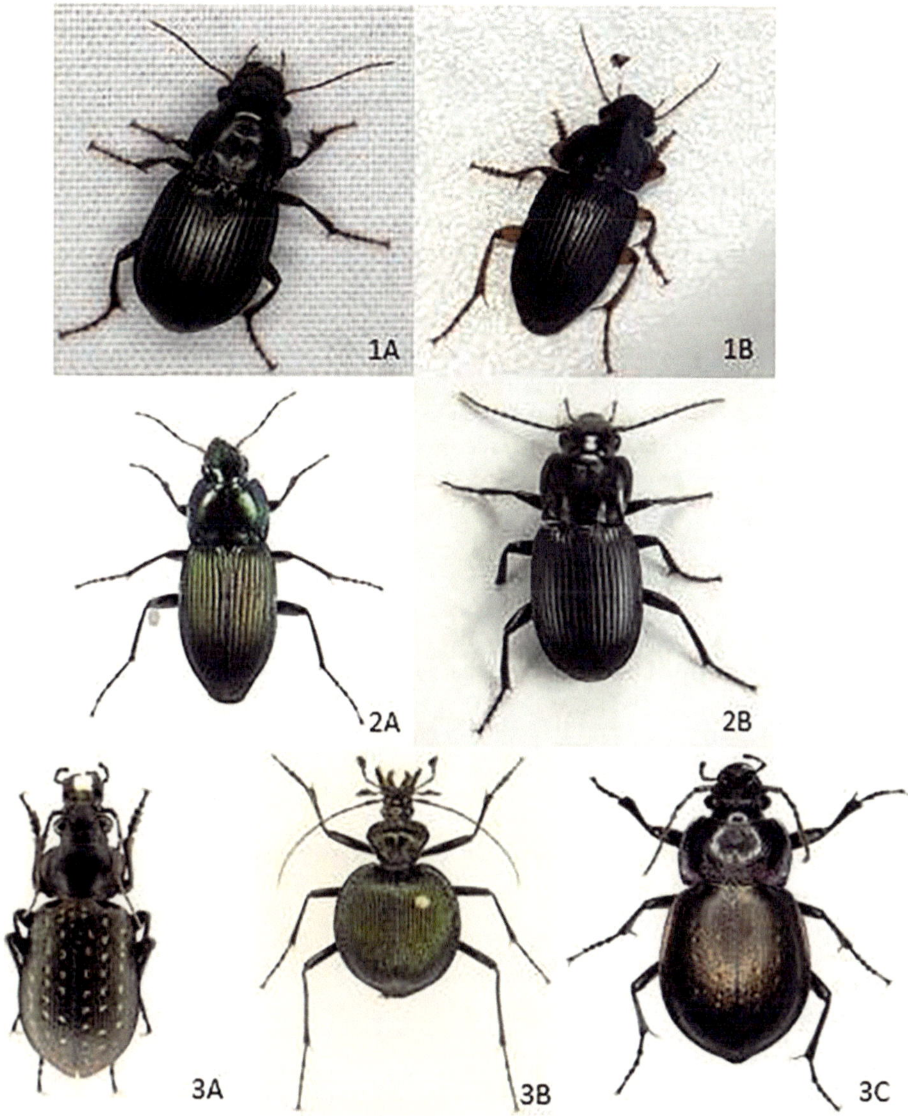

Figure 1.
Based on their feeding habits, Ground beetles are categorized into 1. Granivores – 1A. Amara, 1B. Harpulus, 2. Omnivores – 2A. Poecilus 2B. Pterostichus, 3. Carnivores – 3A. Calosoma calidum, 3B. Scaphinotus, 3C. Carabus nemoralis.

4. Significance of ground beetles

4.1 Ground beetles as predators of insect pests and weed seeds

Ground beetles are effective biocontrol agents for managing insect pests and weed seeds in agroecosystems [11]. Various carabid species like *Anchomenus dorsalis, A. similata, Loricera pilicornis,* and *Premnotrypes suturicallus* play crucial roles in managing pests such as aphids and potato weevils [12]. Additionally, *Trirammatus aerea,* has shown a strong appetite for aphid pests in alfalfa and *Medicago sativa* [13]. Several carabid species from genera, *Galerita, Scarites,* and *Tetracha* may feed on larvae of *Anticarsia gemmatalis* in soybean and maize fields [14]. Ground beetles have also been effective in controlling the fall armyworm in maize [15]. In summary, carabid beetles are valuable allies for controlling agricultural pests. Further details about other important predatory carabids and their prey can be found in **Table 1**.

While ground beetles are known for their role as generalist predators in various agroecosystems, their contribution to weed control is often overlooked [27]. The behavior of carabid beetles as granivores was first described by Forbes in 1883. Carabid species, like *Harpalus affinis* and *Pseudophonus rufipes* may feed on seeds of Canada thistle. Other carabids such as *Ophonus puncticeps* and *O. ardosiacus* larvae, exclusively feed on umbelliferous seeds [28]. Additionally, the genera *Amara, Anisodactylus, Harpalus,* and *Stenolophus* have been shown to reduce seed production in weeds like lambsquarters and redroot pigweed by feeding on them [29]. Occasionally, ground beetles act as a pest of crop plants. For instance, the carabid species *Harpalus smaragdinus* can cause damage to cereals and feed on ripening grains [30].

4.2 Ground beetles as bio-indicators

Bioindicators are organisms that provide valuable insights into ecosystem health and quality. One group of bioindicators that has achieved increased attention in

Carabid beetle	Prey insect/weed	Reference
Loricera pilicornis	Collembolans	[1]
Calosoma granulatum	*Alabama argillacea, Spodoptera frugiperda*	[14]
Scarites orientalis	*Slugs, Sarasinula sp.*	[16]
Scarites anthracinus	*Deroceras reticulatum*	[17]
Chlaenius micans	*Plutella xylostella*	[18]
Dolichus halensis	*P. xylostella*	[18]
Chlaenius Chinese	*Mamestra brassicae*	[19]
Trirammatus aerea	Aphids	[20]
Bembidion quadrimaculatum	*Agrotis ipsilon* and red cut worm	[20]
Pterostichus melanarius	Aphids, carrot weevil, bean weevil, codling moth, Onion fly pupae	[21]
Chlaenius panagaeoides	*Aphis craccivora*	[22]
Casnoidea viridis	*Lamprosema sp.* *Sylepta derogata*	[23]
Neoaulacoryssus cupripennis	*Amaranthus quitensis*	[24]
Harpalus pensylvanicus	*Ambrosia trifida* *Setaria faberi*	[14, 25]
Harpalus rufipes	*Amaranthus retroflexus*	[26]

Table 1.
Carabid beetles in biological control of pests and weeds.

ecological research is ground beetles. These beetles have a potential as a biodiversity indicator [20, 21] since studies have revealed a positive correlation between the richness of carabid species and other beetle families such as Scarabaeidae and Pselaphidae [22]. Carabids have been widely and successfully used as environmental indicators, particularly for assessing environmental pollutants, forest fragmentation [21], and management practices [23]. Furthermore, they have played a pivotal role in studies related to urban ecology [24], the effect of insecticides [25], the impact of military tanks [26], habitat classification, and site quality assessment [31].

Carabid beetles are commonly chosen for eco-toxicological studies because they can be easily found in a wide range of terrestrial habitats [32]. Among them, *Parallelomorphus laevigatus* have been used as an indicator of toxic elements, while *Chlaenius olivieri* serves as an indicator of heavy metal pollution [33]. High bioaccumulation of arsenic (As) and mercury (Hg) has been studied in *Carabus lefebvrei,* and of copper and zinc in *Pterostichus oblongopunctatus,* making them a preferred choice for assessing metal contamination [34].

4.3 Ground beetles as ecosystem engineers

Ground beetles can be considered ecosystem engineers due to their significant impact on their surrounding environments. They play vital roles in shaping and maintaining ecosystems through various means. The key contribution is regulating predator-prey dynamics, which reduces the reliance on chemical pesticides in agriculture. Additionally, they are involved in seed predation and dispersal, which contributes to the diversity of plants. Many soil-dwelling ground beetles create pits in the soil, which improves soil structure and facilitates aeration and water infiltration. These pits also provide shelter to other soil organisms. Furthermore, different development stages of ground beetles feed on litter and dead plant material, thereby promoting nutrient recycling, the decomposition process, and improving soil health [35].

4.4 Ground beetles as pollinators

The ecological significance of ground beetles lies more in their roles as predators and decomposers in terrestrial ecosystems than pollinators. While they may unintentionally contribute to pollination to some extent, their role as pollinators is relatively negligible compared to specialized pollinator species such as honey bees. The Delta green ground beetle, *Elaphrus viridis* is pollinating fruits and vegetables. *Amara aenea* and *A. similata* are pollinators in white mustard [36].

5. Causes for the decline of carabids

A reduction in the diversity of ground beetles (both in population size and species number) in England, affecting approximately 75% of the existing species, has been recorded between 1994 and 2008 [37]. Similarly, Belgium, Denmark, and the Netherlands have also witnessed declines in ground beetle populations of 46%, 31%, and 42%, respectively, over the past 50 to 100 years [38]. The diminishing numbers of carabid beetles have raised significant concerns in the fields of entomology and ecology. These diminutive but ecologically vital insects are facing numerous challenges that have contributed to their waning populations. The key factors behind the decline of carabids are discussed below.

5.1 Habitat loss, fragmentation and degradation

The primary factor leading to the decline in ground beetle diversity is habitat loss, fragmentation, and degradation [39–42]. Ground beetles are recognized as highly significant meso and macro-invertebrates in deglaciated terrains and glacier environments, both in terms of the number of species and their abundance [43]. Habitat changes such as glacier disappearance or alterations in micro-habitat conditions (permafrost melt) can lead to extinction of endemic cryophilous ground beetles [44]. The rapid climate change represents a major concern for high-altitude carabid beetles, specifically for the species living in glaciated mountain areas. There is evidence of an altitudinal shift of the species *Dyscolus diopsis* along the slopes of the Pichincha volcano in Ecuador [45]. Variation in certain life history traits, area contraction, population fragmentation, and shift toward high altitude makes ground beetles useful "sentinels" of environmental and climate changes in mountain areas.

Forest fragmentation results in alterations of habitat size and structure, leading to species decline and localized extinctions. Additionally, it causes isolation and a reduction in population size. Deforestation causes a decline in the abundance of many species and prompts their emigration. It appears that certain large "forest species" like *Damaster blaptoides* and *Haplochlaenius costiger* may have disappeared due to the separation of forest patches from continuous woodlands, while medium-sized "forest species" like *Carabus yaconinus*, *Diplochelia zeelandica*, and *Brachinus scotomedes* are diminishing as forested areas shrink. *Carabus scheidleri* and *Carabus coriaceus* are dominant in the oldest forests [46] but deforestation leads to adverse effects on microclimate conditions, substantial damage to the herbaceous cover, and disruption of the surface soil [46]. Furthermore, the extent of grazing also had an impact on ground beetle populations and community composition [47]. The highest ground beetle numbers were recorded in areas subjected to less intensive grazing [48].

5.2 Intensive use of pesticides

The use of pesticides, particularly broad-spectrum insecticides has adverse effects on carabid beetles [49, 50]. They exhibit varying degrees of susceptibility to different pesticides across species and different life stages [51]. Some species are tolerant to pyrethroids, while others are sensitive to neonicotinoids [52]. These pesticides act on ground beetles in two ways, either by killing the beetles through contact or consumption, or indirectly poisoning the prey with insecticides, causing lethal effects to carabid beetles and consequently death [53]. Insecticides also show sub-lethal effects by interfering with their reproductive development or behavior, thereby causing a decrease in the number of offspring [53]. For instance, the combined use of conventional insecticides and neonicotinoid-treated seeds results in high mortality of *Harpalus pensylvanicus* [52].

5.3 Agronomic practices

The major agronomic practice that affects the carabids is tillage. It affects ground beetles in several ways, by influencing the physical properties of the soil [54], water balance [55], and aeration as well as the distribution of plant residues in the tilled layer [56]. Reduced ground cover also makes the soil and the associated ground beetle populations more vulnerable to climate conditions [57, 58]. The heavy machinery used for tillage exerts strong forces that can physically kill ground beetles [59] or expose soft-bodied larval instars to the surface and make ground beetles easy for predation [60]. Reduced tillage intensity is associated with a greater variety of carabid species, indicating that tillage affects both population numbers and species composition [61].

5.4 Climate change

Insect communities seem to react faster to temperature fluctuations than vegetation. Subterranean carabid beetles provide valuable insights into the effects of global warming because they exhibit sensitivity to temperature shifts and alterations in environmental conditions [62]. The influence of climate change is often overshadowed by human activities and placing them at risk. This is exemplified by the gradual extinction of three species such as *Pterostichus burmeisteri, Pterostichus unctulatus, and Carabus auronitens* in specific parts of their habitats as a result of transformations [62].

The consequences of global warming on carabid species, particularly those with limited mobility or predominantly inhabiting forested species, pose a significant threat to their survival [62]. Moreover, it also induces alterations in the habitats and behaviors of hypogean carabid beetles, causing them to move toward the earth's surface. Furthermore, rising temperatures associated with global warming have led to the discovery of previously unknown carabid subgroups in the subterranean environment, such as *Promecognathini and Lovriciina* [62].

5.5 Invasive aliens

Invasive aliens are non-native organisms that are introduced into environments beyond their natural habitat, often causing detrimental effects on the native biodiversity. Invasive European carabid beetles have been linked to population declines in native species. For instance, *Pterostichus melanarius* appears to have an inverse relationship with the population of *Pterostichus adstrictus* [63]. Similarly, the exotic carabid beetle, *Merizodus soledadinus*, has spread to various regions by preying on invertebrates. Its expanded geographical distribution could have significant repercussions on the abundance of its prey species. Invasive species that occupy higher positions in the food chain often exert substantial ecological impacts, potentially leading to reduced biodiversity and alterations in the dynamics of trophic interactions.

5.6 Light pollution

Light pollution poses a significant threat to both insects and biodiversity [64]. Ground beetles are known to exhibit positive phototaxis for light [65]. Insects that are attracted to vehicle headlights often meet an untimely death. This "fatal attraction" results in the death of roughly 100 billion insects [66]. When insects are attracted to artificial light sources, their lives typically perish either due to exhaustion or predation [67].

The impact of artificial light extends across various facets of ground beetles such as behavior, navigation, competitive interactions, predator-prey dynamics, and physiology. The attractive and repulsive effects of artificial light disrupt the natural patterns of movement [68] and lead to alteration s in insect distribution within landscapes, diverging from their evolutionary baseline [69–71]. Artificial light at night functions as an ecological trap, accumulating insects in higher numbers and diminishing the fitness of nocturnal ground beetles that are attracted to them [72]. Artificial light at night also creates a false impression of an increase in a species' population, even when species are declining in areas outside the illuminated zones. This indicates that predatory ground beetles may be more distracted by light than exploiting it [73].

6. Conservation strategies for ground beetles

Ground beetles play a crucial role in maintaining ecosystem health by preying on pests and managing insect populations in both natural and agricultural ecosystem. Their predatory behavior contributes significantly to ecosystem stability and biodiversity. Nevertheless, like many species, ground beetles face multiple threats, resulting in population declines. To ensure the continued existence of ground beetles and the valuable ecological services they provide, effective conservation measures are essential.

The decline of ground beetles can be attributed to a complex interplay of various factors. It is crucial to consider the multifaceted nature of the problem and implement conservation efforts to safeguard ground beetles and contribute to the broader conservation of biodiversity.

The holistic approaches to the conservation of ground beetles are given below.

6.1 Habitat management

Habitat management plays a critical role in conserving ground beetles [74]. It involves the deliberate manipulation and maintenance of natural and anthropogenic habitats to create favorable conditions for ground beetles. Habitat management for ground beetles is divided into non-crop refuge habitats and beetle banks.

6.1.1 Non-crop refuge habitats

Ground beetle habitat in permanent vegetation can be strategically located at field edges, marginal lands, or specific areas within crop fields, making them excellent sites for ground beetle conservation. A beneficial approach to enhance ground beetle populations involves maintaining non-crop "refuge" habitats. These refuge areas, often found at field edges or within fields where minimal management occurs, offer ground beetles to access their prey, including insect pests and weed seeds. The refuge habitats are valuable during disturbance events like pesticide application and pruning, as they facilitate the recolonization of crop areas by ground beetles. Undisturbed mature ecosystems, particularly nature reserves [75, 76], edge habitats, and habitat mosaics [77], establishing tall grass areas and the use of living mulches such as red clover and white clover between crop rows creates an ideal microclimate and boosts ground beetle abundance and diversity [78, 79]. *Carabus hungaricus* can be protected by active management of open meadows.

Regeneration of forest field mosaics encouraged new specimens of *Carabus auronitens* due to genetic flow from the forest remnants, leading to increased diversity [80]. Hedgerows also help in the conservation of ground beetles as they provide a habitat for recolonization. The significance of hedgerows in supporting forest carabid communities has been emphasized [81]. Roadside verges could serve both as habitats and as pathways connecting different heathland patches, which maintains the population of the carabids [82]. This dual role of roadsides was also documented in the context of Finnish forest roads [83] and in relation to highway verges in the Netherlands [84]. Therefore, the remaining fragments of natural habitat, the corridors within similar, often human-made environments, and the larger, artificially created patches, may collectively establish an effective network for the conservation of carabid beetles [85].

The conservation value for threatened ground beetles were studied in the small meadows of forest-dominated areas, which could serve as crucial refuge habitats for many invertebrates, particularly ground beetles. Fifty-two carabid species occupied different habitat preferences in the meadows. Smaller meadows had higher carabid species richness, as they attracted forest-inhabiting species. Forests influenced the number of open-habitat species, indicating the modest conservation value of these meadows for ground beetles in a forested landscape [86]. The grasslands mowed annually and those under alternative management of both grazing and mowing had greater populations of ground beetles and a wider variety of species compared to areas subjected to grazing. This indicates that engaging simplified, organic farming approaches involving reduced mowing and a combination of grazing can effectively protect and maintain biodiversity in mountainous region grasslands [48].

6.1.2 Beetle banks

Beetle banks are a widely adopted conservation technique, especially popular in Europe, involving strips of land approximately two yards wide adjacent to crop fields. These strips are planted with specific grasses and perennial flowers to create overwintering and refuge habitats [87, 88]. This approach is a form of biological pest control, aimed at increasing the population of natural enemies to check the pests in crops. Beetle banks serve as a habitat for these natural enemies, leading to an increase in predator populations particularly ground beetles, and a reduction in pest populations. The grasses and weeds used in beetle banks were *Setaria sp., Trifolium pretense,* and *Taraxacum officinale,* has the added benefit of attracting ground beetles to crop fields [89]. Notably, ground beetle species like *Pseudophoonus rufipes* and *Harpalus differendus* are commonly found inhabitants of beetle banks [90]. In the fields adjacent to beetle banks and other uncultivated areas, ground beetles are twice as abundant compared to cultivated fields [91].

6.2 Use of no pesticides/insecticides

The establishment of beetle banks, as previously explained, stands as the most effective method to counteract the impact of insecticides. The utilization of narrow-spectrum pesticides or insecticides represents a precision-focused strategy aimed at preserving ground beetles and other beneficial insects. The narrow spectrum insecticides such as Bt (*Bacillus thuringiensis*) or insect growth regulators such as Methoxyfenozide (Intrepid®) have little or no effect on ground beetles [92]. The fungal pathogen *Beauveria bassiana* (Botanigard®) has been tested against the ground beetle *Harpalus rufipes* and has been found to cause little to no mortality [92]. These pesticides contribute to more sustainable and ecologically friendly pest management practices, helping to protect the valuable ecosystem services provided by ground beetles.

6.3 Sustainable farm management practices

The key to harnessing the advantages of carabid beetles in agriculture lies in enhancing their survival rates. To achieve this, specific agricultural practices can be implemented to conserve carabid populations like reduced tillage. Non-inversion tillage minimizes harm to eggs, larvae, and adult carabid beetles compared to inversion tillage [58]. Moreover, minimum tillage also play a role in conserving surface vegetation and mulch, creating microhabitats that provide shelter for ground beetles. These microenvironments offer protection from adverse environmental conditions and potential predators. The adoption of practices such as promoting surface residues [93], embracing non-inversion tillage methods, and selecting appropriate mowing techniques [94] can collectively contribute to the conservation and flourishing of ground beetles in the agricultural ecosystems.

Furthermore, the use of compost around plants has been shown to enhance carabid beetle abundance in comparison to bare soil [95]. Additionally, Organic farming practices, particularly those involving reduced pesticide usage and fewer chemical inputs, tend to support higher levels of activity and biodiversity among carabid beetles in contrast to conventionally managed fields [65, 96]. These results underscore the importance of adopting ecologically mindful practices to foster the well-being of carabid beetles.

6.4 Reduce light pollution

While contemporary living demands artificial light designs and embellishments, it's imperative to remain mindful of the adverse effects of light on insects. To address these, it is advisable to promote awareness among the public through training, workshops, social media, or collaboration with government agencies, encouraging minimal use of artificial lighting or even discontinuation of unnecessary lighting systems. Furthermore, state or national governments should partner with lighting experts to explore and create lighting systems that are conducive to insect well-being, ensuring the preservation of future generations of insects without the need for insect trapping [97].

6.5 Monitoring

Monitoring ground beetles for conservation is a vital component of safeguarding biodiversity. This involves a specific focus on a particular species, the evaluation of habitat quality, or the tracking of environmental shifts of ground beetles. Notably, the Globe Net project has monitored and documented, an increased prevalence of specific dominant species within urban forests [98–100].

The Environmental Change Network (ECN) stands as a longstanding monitoring project that was founded in 1993. Under the ECN umbrella, the Ground Predator Protocol, with a primary focus on ground beetles, has been actively conducted at two ECN sites in Scotland, specifically at Glensaugh and Sourhope, since 1994. This ongoing effort has yielded a continuous dataset that covers 18 years. Consequently, this monitoring initiative contributes to the development of conservation strategies aimed at protecting ground beetles.

6.6 Citizen science intervention

Citizen science involves the participation of the general public in scientific undertakings [101]. The Current progress in information technology has created new opportunities for citizen science projects to engage a large number of individuals in monitoring the natural environment and biodiversity across extensive geographical areas [102, 103]. Involving citizen scientists in initiatives dedicated to preserving ground beetles can significantly enhance data collection, generate awareness regarding the importance of ground beetles, and nurture a sense of responsibility and care for local ecosystems. This approach allows the public to participate in meaningful scientific research and gather valuable conservation data. The data collected through citizen science efforts can be transformed into practical conservation strategies, including habitat restoration, pest management strategies, and recommendations for policies.

7. Conclusion

Ground beetles are incredibly diverse, with over 40,000 species worldwide, found across a range of ecosystems from tropical rainforests to deserts. They vary in size, shape, coloration, and behavior, occupying different ecological niches. Carabid beetles can be categorized based on their habitat preferences as generalists or specialists. They are also classified by their feeding habits, including granivores, omnivores, and carnivores. They are essential as they act as effective biocontrol agents, managing insect pests and weed seeds in agroecosystems. They are also used as bioindicators to assess environmental health and quality and play roles as ecosystem engineers and, to a lesser extent, as pollinators. Several factors contribute to the decline of ground beetles, including habitat loss, fragmentation, and degradation, intensive pesticide use, agronomic practices like tillage, climate change, invasive species, and light pollution. Conservation strategies for ground beetles include habitat management through non-crop refuges and beetle banks, the use of narrow-spectrum pesticides, sustainable farm management practices, reducing light pollution, monitoring populations, and involving citizen scientists in data collection and awareness efforts.

Author details

Bindu Gudi Ramakrishna*, Jasti Sri Vishnu Murthy, Gopu Sushma and C.N. Thanu
Department of Agricultural Entomology, College of Agriculture, Thrissur, India

*Address all correspondence to: bindugudi22@gmail.com

References

[1] Kromp B. Carabid beetles in sustainable agriculture: A review on pest control efficacy, cultivation impacts and enhancement. Agriculture, Ecosystems & Environment. 1999;**74**(1-3):187-228. DOI: 10.1016/S0167-8809(99)00037-7

[2] Larochelle A. The Food of Carabid Beetles (Coleoptera: Carabidae, Including Cicindelinae) Fabreries, supplément 5. Sillery: Association des Entomolgistes Amateurs du Québec, Quebec City, Quebec, Canada; 1990

[3] Lövei GL, Sunderland KD. Ecology and behavior of ground beetles (Coleoptera: Carabidae). Annual Review of Entomology. 1996;**41**(1):231-256

[4] Brandmayr P. Natural history and applied ecology of carabid beetles. In: Proceedings of the IXth European Carabidologists' Meeting, 26-31 July 1998, Camigliatello, Cosenza, Italy. Vol. 19. Bulgaria: Pensoft Publishers; 2000

[5] Rivard I. Ground beetles (Coleoptera: Carabidae) in relation to agricultural crops. The Canadian Entomologist. 1966;**98**(2):189-195

[6] Tomlin AD. The toxicity of insecticides by contact and soil treatment to two species of ground beetles (Coleoptera: Carabidae) 1. The Canadian Entomologist. 1975;**107**(5):529-532

[7] Lund RD, Turpin FT. Carabid damage to weed seeds found in Indiana cornfields. Environmental Entomology. 1977;**6**(5):695-698

[8] Allen RT. The occurrence and importance of ground beetles in agricultural and surrounding habitats. In: Carabid Beetles: Their Evolution, Natural History, and Classification. Dordrecht: Springer Netherlands; 1979. pp. 485-505

[9] Morrill WL. Ground beetles (Coleoptera: Carabidae) in Georgia grasslands. Journal of Agricultural Entomology. 1992;**9**(3):179-188

[10] Eversham BC, Roy DB, Telfer MG. Urban, industrial, and other manmade sites as analogues of natural habitats for Carabidae. In: Annales Zoologici Fennici. Finland: Finnish Zoological and Botanical Publishing Board; 1996. pp. 149-156

[11] Sunderland KD. Invertebrate pest control by carabids. In: Holland JM, editor. The Agroecology of Carabid Beetles. Andover, Hampshire, UK: Intercept Ltd; 2002. pp. 215-231

[12] Symondson WO. Molecular identification of prey in predator diets. Molecular Ecology. 2002;**11**(4):627-641

[13] Grez AA, Rivera P, Zaviezo T. Foliar and ground-foraging predators of aphids associated with alfalfa crops in Chile: Are they good or bad partners? Biocontrol Science and Technology. 2007;**17**(10):1071-1077. DOI: 10.1080/09583150701748146

[14] Cividanes FJ. Carabid beetles (Coleoptera: Carabidae) and biological control of agricultural pests in Latin America. Annals of the Entomological Society of America. 2021;**114**(2):175-191. DOI: 10.1093/aesa/saaa051

[15] Wyckhuys KAG, O'Neil RJ. Influence of extra-field characteristics to abundance of key natural enemies of Spodoptera frugiperda Smith (Lepidoptera: Noctuidae) in subsistence maize production. International Journal

of Pest Management. 2007;**53**(2):89-99. DOI: 10.1080/09670870701245207

[16] Bennett FD, Yaseen M. Investigations for the possibilities of biological control of slugs in Honduras. Progress Report: CEIBA. 1987;**28**:229-234

[17] Tulli MC, Carmona DM, Lopez AN, Manetti PL, Vincini AM, Cendoya G. Predación de la babosa Deroceras reticulatum (Pulmonata: Stylommathophora) por Scarites anthracinus (Coleoptera: Carabidae). Ecología Austral. 2009;**19**(1):55-61

[18] Yamada H. Food consumption of four predators of the diamondback moth, Plutella xylostella (L.), Chlaeninus (Chlaenius) micans (Fabricius), Paederus fuscipes Curtis, Philonthus wusthoffi Bernk and Labidura riparia (Pallas). Japan Journal of Applied Entomology and Zoology. 1985;**29**:173-175

[19] Kuwayama S, Oshima K. Ecological Studies on Calosoma Chinense, a Predacious Carabid against Army and Cut-Worms, and Some Related Species. Sapporo, Canada; 1964. p. 46

[20] Frank JH. Carabidae (Coleoptera) as predators of the red-backed cutworm (Lepidoptera: Noctuidae) in Central Alberta. Canadian Entomologist. 1971;**103**(7):1039-1044. DOI: 10.4039/Ent1031039-7

[21] Swaminathan R. Ground beetles (Coleoptera: Carabidae): Their potential as bio-agents in agroecosystems. Basic and Applied Aspects of Biopesticides. 2014:225-233

[22] Rajagopal D, Kumar P. Predation potentiality of *Chlaenius panagaeoides* (Laferte) (Coleoptera: Carabidae) on cowpea aphid *Aphis craccivora* Koch (Homoptera: Aphididae). Journal of Aphidology. 1988;**2**:93-99

[23] Swaminathan R, Bhati KK, Hussain T. Preliminary investigations on the predation potential of carabids. Indian Journal of Applied Entomology. 2001;**15**:37-41

[24] Nisensohn L, Faccini D, Montero G, Lietti M. Predation of *Amaranthus quitensis* HBK seeds in soybean crops: Influence of the tillage system. Pesquisa Agropecuaria Brasileira. 1999;**34**(3):377-384. DOI: 10.1590/S0100-204X1999000300008

[25] Ward MJ, Ryan MR, Curran WS, Barbercheck ME, Mortensen DA. Cover crops and disturbance influence activity density of weed seed predators *Amara aenea* and *Harpalus pensylvanicus* (Coleoptera: Carabidae). Weed Science. 2011;**59**(1):76-81. DOI: 10.1614/WS-D-10-00065.1

[26] Harrison S, Gallandt ER. Behavioural studies of *Harpalus rufipes* De Geer: An important weed seed predator in northeastern US agroecosystems. International Journal of Ecology. 2012:6. DOI: 10.1155/2012/846546

[27] Kulkarni SS, Dosdall LM, Willenborg CJ. The role of ground beetles (Coleoptera: Carabidae) in weed seed consumption: A review. Weed Science. 2015;**63**(2):355-376

[28] Zetto-Brandmayr T. Life cycle, control of propagation rhythm and fecundity of Ophonus rotundicollis Fairm et Lab (Coleoptera, Carabidae, Harpalini) as an adaptation to the mainfeeding plant Daucus carota L. Brandmayr P., den Boer P.J., Weber F. (eds.). The Synthesis of Field Studies and Laboratory Experiment. Report of the 4th Meeting of European Carabidologists, 'Haus Rothenberge', Westphalia Bundesrepublik Deutschland. Wageningen: Centre for Agricultural Publishing and Documentation (PUDOC). 1983. pp. 93-103

[29] Harrison SK, Regnier EE, Schmoll JT. Postdispersal predation of giant ragweed (Ambrosia trifida) seed in no-tillage corn. Weed Science. 2003;**51**(6):955-964. DOI: 10.1614/P2002-110

[30] Brust GE. Seed-predators reduce broadleaf weed growth and competitive ability. Agriculture, Ecosystems and Environment. 1994;**48**:27-34. DOI: 10.1016/0167-8809(94)90072-8

[31] Duelli P, Obrist MK. In search of the best correlates for local organismal biodiversity in cultivated areas. Biodiversity and Conservation. 1998;**7**:297-309

[32] Niemela J, Baur B. Threatened species in a vanishing habitat: Plants and invertebrates in calcareous grasslands in the Swiss Jura Mountains. Biodiversity and Conservation. 1998;**7**:1407-1416

[33] Oliver I, Beattie AJ. Designing a cost-effective invertebrate survey. A test of methods for rapid assessment of biodiversity. Ecological Applications. 1996;**6**:594-607. DOI: 10.2307/2269394

[34] Rushton SP, Eyre MD, Luff ML. The effects of management on the occurrence of some ground beetle species in grassland. In: Stork NE, editor. The Role of Ground Beetles in Ecological and Environmental Studies. Andover, UK: Intercept; 1990. pp. 209-216

[35] Venn S. The Effects of Urbanization on Boreal Forest Ecosystems [thesis]. Helsinki, Finland: University of Helsinki; 2000

[36] Naumkin VP, Velkova NI. Species diversity of insects-pollinators on crops of white mustard. Вестник аграрной науки. 2013;**43**(4):28-32

[37] Brooks DR, Bater JE, Clark SJ, Monteith DT, Andrews C, Corbett SJ, et al. Large carabid beetle declines in United Kingdom monitoring network increases evidence for a widespread loss in insect biodiversity. Journal of Applied Ecology. 2012;**49**:1009-1019. DOI: 10.1111/j.1365-2664.2012.02194.x

[38] Kotze DJ, O'hara RB. Species decline—But why? Explanations of carabid beetle (Coleoptera, Carabidae) declines in Europe. Oecologia. 2003;**135**:138-148. DOI: 10.1007/s00442-002- 1174-3

[39] De Vries HH. Size of habitat and presence of ground beetle species. In: Carabid Beetles: Ecology and Evolution. Dordrecht: Springer Netherlands; 1994. pp. 253-259

[40] Keller I, Excoffier L, Largiader CR. Estimation of effective population size and detection of a recent population decline coinciding with habitat fragmentation in a ground beetle. Journal of Evolutionary Biology. 2005;**18**(1):90-100. DOI: 10.1111/j.1420-9101.2004.00794.x

[41] Hendrickx F, Maelfait JP, Desender K, Aviron S, Bailey D, Diekotter T, et al. Pervasive effects of dispersal limitation on within-and among-community species richness in agricultural landscapes. Global Ecology and Biogeography. 2009;**18**(5):607-616. DOI: 10.1111/j.1466-8238.2009.00473.x

[42] Massaloux D, Sarrazin B, Roume A, Tolon V, Wezel A. Complementarity of grasslands and cereal fields ensures carabid regional diversity in French farmlands. Biodiversity and Conservation. 2020;**29**:2861-2882

[43] Gobbi M, Lencioni V. Glacial Biodiversity: Lessons from grounddwelling and aquatic insects. 2020:1-23

[44] Gobbi M. Global warning: challenges, threats and opportunities for ground beetles (Coleoptera: Carabidae) in high altitude habitats. arXiv preprint arXiv:2011.06804. 2020

[45] Moret P, Aráuz MD, Gobbi M, Barragán Á. Climate warming effects in the tropical Andes: First evidence for upslope shifts of Carabidae (Coleoptera) in Ecuador. Insect Conservation and Diversity. 2016;**9**(4):342-350

[46] Fujita A, Maeto K, Kagawa Y, Ito N. Effects of forest fragmentation on species richness and composition of ground beetles (Coleoptera: Carabidae and Brachinidae) in urban landscapes. Entomological Science. 2008;**11**(1):39-48

[47] Gardner SM, Hartley SE, Davies A, Palmer SCF. Carabid communities on heather moorlands in Northeast Scotland: The consequences of grazing pressure for community diversity. Biological Conservation. 1997;**81**(3):275-286. DOI: 10.1016/S0006-3207(96)00148-6

[48] Twardowski JP, Pastuszko K, Hurej M, Gruss I. Effect of different management practices on ground beetle (Coleoptera: Carabidae) assemblages of uphill grasslands. Polish Journal of Ecology. 2017;**65**(3):400-409. DOI: 10.3161/15052249PJE2017.65.3.007

[49] Van Toor RF. The effects of pesticides on carabidae (Insecta: Coleoptera), predators of slugs (Mollusca: Gastropoda): Literature review. New Zealand Plant Protection. 2006;**59**:208. DOI: 10.30843/nzpp.2006.59.4543

[50] Alam MJ, Das G. Toxicity of insecticides to predators of rice brown planthopper: Wolf spider and carabid beetle. Journal of Food, Nutrition and Agriculture. 2020;**3**:9-13. DOI: 10.21839/jfna.2020.v3.310

[51] Leslie TW, Biddinger DJ, Mullin CA, Fleischer SJ. Carabidae population dynamics and temporal partitioning: Response to coupled neonicotinoid-transgenic technologies in maize. Environmental Entomology. 2009;**38**(3):935-943. DOI: 10.1603/022.038.0348

[52] Molnar I, Rakosy-Tican E. Difficulties in potato pest control: The case of pyrethroids on Colorado potato beetle. Agronomy. 2021;**11**(10):1920

[53] Mulligan EA, Ferry N, Jouanin L, Walters KF, Port GR, Gatehouse AM. Comparing the impact of conventional pesticide and use of a transgenic pest-resistant crop on the beneficial carabid beetle *Pterostichus melanarius*. Pest Management Science: Formerly Pesticide Science. 2006;**62**(10):999-1012

[54] Nunes MR, Karlen DL, Veum KS, Moorman TB, Cambardella CA. Biological soil health indicators respond to tillage intensity: A US meta-analysis. Geoderma. 2020;**369**:114335. DOI: 10.1016/j.geoderma.2020.114335

[55] Brunel-Saldias N, Seguel O, Ovalle C, Acevedo E, Martínez I. Tillage effects on the soil water balance and the use of water by oats and wheat in a Mediterranean climate. Soil and Tillage Research. 2018;**184**:68-77. DOI: 10.1016/j.still.2018.07.005

[56] Seitz S, Goebes P, Puerta VL, Pereira EIP, Wittwer R, Six J, et al. Conservation tillage and organic farming reduce soil erosion. Agronomy for Sustainable Development. 2019;**39**:1-10

[57] Larochelle A, Larivière MC. A Natural History of the Ground-Beetles

(Coleoptera: Carabidae) of America North of Mexico. Sofia, Bulgaria: Pensoft; 2003. p. 583

[58] Shearin AF, Reberg-Horton SC, Gallandt ER. Direct effects of tillage on the activity density of ground beetle (Coleoptera: Carabidae) weed seed predators. Environmental Entomology. 2014;**36**(5):1140-1146. DOI: 10.1603/0046-225X(2007)36[1140:DEOTOT]2.0.CO;2

[59] Fadl A, Purvis G, Towey K. The effect of time of soil cultivation on the incidence of Pterostichus melanarius (Illig.) (Coleoptera: Carabidae) in arable land in Ireland. In: Annales Zoologici Fennici. Finnish Zoological and Botanical Publishing Board; 1996;**33**:207-214

[60] Thorbek P, Bilde T. Reduced numbers of generalist arthropod predators after crop management. Journal of Applied Ecology. 2004;**41**(3):526-538. DOI: 10.1111/j.0021-8901.2004.00913.x

[61] Kennedy TF, Connery J, Fortune T, Forristal D, Grant J. A comparison of the effects of minimum-till and conventional-till methods, with and without straw incorporation, on slugs, slug damage, earthworms and carabid beetles in autumn-sown cereals. The Journal of Agricultural Science. 2013;**151**(5):605-629. DOI: 10.1017/S0021859612000706

[62] Brandmayr P, Giorgi F, Casale A, Colombetta G, Mariotti L, Taglianti AV, et al. Hypogean carabid beetles as indicators of global warming? Environmental Research Letters. 2013;**8**(4):044047

[63] Spence JR, Spence DH. Of ground-beetles and men: Introduced species and the synanthropic fauna of western Canada. The Memoirs of the Entomological Society of Canada. 1988;**120**(S144):151-168

[64] Grubisic M, van Grunsven RH, Kyba CC, Manfrin A, Hölker F. Insect declines and agroecosystems: Does light pollution matter? Annals of Applied Biology. 2018;**173**(2):180-189. DOI: 10.1111/aab.12440

[65] Owens AC, Cochard P, Durrant J, Farnworth B, Perkin EK, Seymoure B. Light pollution is a driver of insect declines. Biological Conservation. 2020;**241**:108259. DOI: 10.1016/j.biocon.2019.108259

[66] Gaston KJ, Holt LA. Nature, extent and ecological implications of night-time light from road vehicles. Journal of Applied Ecology. 2018;**55**(5):2296-2307. DOI: 10.1111/1365-2664.13157

[67] Eisenbeis G, Rich C, Longcore T. Artificial night lighting and insects: Attraction of insects to streetlamps in a rural setting in Germany. Ecological Consequences of Artificial Night Lighting. 2006;**2**:191-198

[68] Allema AB, Rossing WAH, Van der Werf W, Heusinkveld BG, Bukovinszky T, Steingröver E, et al. Effect of light quality on movement of *Pterostichus melanarius* (Coleoptera: Carabidae). Journal of Applied Entomology. 2012;**136**(10):793-800. DOI: 10.1111/j.1439-0418.2012.01728.x

[69] Degen T, Mitesser O, Perkin EK, Weiß NS, Oehlert M, Mattig E, et al. Street lighting: Sex-independent impacts on moth movement. Journal of Animal Ecology. 2016;**85**(5):1352-1360. DOI: 10.1111/1365-2656.12540

[70] Manfrin A, Singer G, Larsen S, Weiß N, Van Grunsven RH, Weiß NS, et al. Artificial light at night affects organism flux across ecosystem

boundaries and drives community structure in the recipient ecosystem. Frontiers in Environmental Science. 2017;**5**:61. DOI: 10.3389/fenvs.2017.00061

[71] Manríquez PH, Jara ME, Diaz MI, Quijón PA, Widdicombe S, Pulgar J, et al. Artificial light pollution influences behavioral and physiological traits in a keystone predator species, *Concholepas concholepas*. Science of the Total Environment. 2019;**661**:543-552. DOI: 10.1016/j.scitotenv.2019.01.157

[72] Luarte T, Bonta CC, Silva-Rodriguez EA, Quijón PA, Miranda C, Farias AA, et al. Light pollution reduces activity, food consumption and growth rates in a sandy beach invertebrate. Environmental Pollution. 2016;**218**:1147-1153. DOI: 10.1016/j.envpol.2016.08.068

[73] Gaston KJ, Duffy JP, Gaston S, Bennie J, Davies TW. Human alteration of natural light cycles: Causes and ecological consequences. Oecologia. 2014;**176**:917-931

[74] Menalled FD, Gross KL, Hammond M. Weed aboveground and seedbank community responses to agricultural management systems. Ecological Applications. 2001;**11**(6):1586-1601. DOI: 10.1890/1051-0761(2001)011[1586:WAASCR]2.0.CO;2

[75] Desender K, Small E, Gaublomme E, Verdyck P. Rural–urban gradients and the population genetic structure of woodland ground beetles. Conservation Genetics. 2005;**6**:51-62

[76] Skłodowski J. Anthropogenic transformation of ground beetle assemblages (Coleoptera: Carabidae) in Bialowieza Forest, Poland: From primeval forests to managed woodlands of various ages. Entomologica Fennica. 2006;**17**(3):296-314. DOI: 10.33338/ef.84349

[77] Andorkó R, Kadar F. Carabid beetle (Coleoptera: Carabidae) communities in a woodland habitat in Hungary. Entomologica Fennica. 2006;**17**(3):221-228. DOI: 10.33338/ef.84334

[78] Blubaugh CK, Hagler JR, Machtley SA, Kaplan I. Cover crops increase the foraging activity of omnivorous predators in seed patches and facilitate weed biological control. Agriculture, Ecosystems & Environment. 2016;**231**:264-270. DOI: 10.1016/j.agee.2016.06.045

[79] Hajek AE, Eilenberg J. Natural Enemies: An Introduction to Biological Control. Cambridge: Cambridge University Press; 2018

[80] Terlutter H. An allele gradient of an esterase gene locus as result of recent gene flow: Electrophoretic investigations of *Carabus auronitens* F.(Col. Carabidae). In: The Role of Ground Beetles in Ecological and Environmental Studies. America: Intercept; 1990. pp. 359-364

[81] Petit S. Diffusion of forest carabid beetles in hedgerow network landscapes. Carabid Beetles: Ecology and Evolution. 1994:337-341

[82] Vermeulen R, Opsteeg T. Movements of some carabid beetles in road-side verges. Dispersal in a simulation programme. In: Carabid Beetles: Ecology and Evolution. Dordrecht: Springer Netherlands; 1994. pp. 393-398

[83] Koivula M. The Forest Road Network-a Landscape Element Affecting the Distribution of Boreal Carabid Beetles (Coleoptera, Carabidae). How to Protect or What We know About Carabid Beetles: from Knowledge to Application, from Wijster (1969) to Tuczno. Poland:

Warsaw Agricultural University Press; (2001)2002. pp. 287-299

[84] Noordijk J. Arthropods in Linear Elements: Occurence, Behaviour and Conservation Management. Netherlands: Wageningen University and Research; 2009

[85] Vermeulen R, Spee A. The Mantingerveld: effects of fragmentation and defragmentation followed by carabid beetles. In: European Carabidology 2003. Proceedings of the 11th European Carabidologists' Meeting, 2005 January. DIAS Report Plant Production. DIAS publications; 2005. pp. 379-389

[86] Buse J, Eckerter T, Eichenseer P, Förschler MI, Oelmann Y, Georgi M. Conservation value of small meadows in a forest-dominated landscape assessed for ground beetles (Coleoptera: Carabidae). Angewandte Carabidologie. 2018;**12**:49-56

[87] Chiverton PA. The creation of within-field overwintering sites for natural enemies of cereal aphids. In: British Crop Protection Conference. UK; 1989;**3**:1093-1096

[88] Collins KL, Boatman ND, Wilcox A, Holland JM, Chaney K. Influence of beetle banks on cereal aphid predation in winter wheat. Agriculture, Ecosystems & Environment. 2002;**93**(1-3):337-350

[89] Karindah S, Purwaningsih A, Agustin A. Ketertarikan Anaxipha longipennis Serville (Orthoptera: Gryllidae) terhadap beberapa jenis gulma di sawah sebagai tempat bertelur. Jurnal Entomologi Indonesia. 2011;**8**(1):27

[90] Cardarelli E, Bogliani G. Effects of grass management intensity on ground beetle assemblages in rice field banks. Agriculture, Ecosystems & Environment. 2014;**1**(195):120-126

[91] Hance T, Impact of Cultivation and Crop Husbandry Practices. 2002.

[92] Jones A, Birthisel S, Jabbour R, Drummond F, Yarborough D. Beneficial Insect Series 2: Carabidae (Ground Beetles) on Maine Farms. America: Oxford University Press; 2013. p. 04469

[93] Shearin AF, Reberg-Horton SC, Gallandt ER. Cover crop effects on the activity density of the weed seed predator Harpalus rufipes (Coleoptera: Carabidae). Weed Science. 2008;**56**(3):442-450. DOI: 10.1614/WS-07-137.1

[94] Humbert JY, Ghazoul J, Richner N, Walter T. Hay harvesting causes high orthopteran mortality. Agriculture, Ecosystems & Environment. 2010;**139**(4):522-527. DOI: 10.1016/j.agee.2010.09.012

[95] Renkema JM, Lynch DH, Cutler GC, MacKenzie K, Walde SJ. Predation by Pterostichus melanarius (Illiger) (Coleoptera: Carabidae) on immature Rhagoletis mendax Curran (Diptera: Tephritidae) in semi-field and field conditions. Biological Control. 2012;**60**(1):46-53. DOI: 10.1016/j.biocontrol.2011.10.004

[96] Tuck SL, Winqvist C, Mota F, Ahnström J, Turnbull LA, Bengtsson J. Land-useintensityandtheeffectsoforganic farming on biodiversity: A hierarchical meta-analysis. Journal of Applied Ecology. 2014;**51**(3):746-755. DOI: 10.1111/1365-2664.12219

[97] Dar SA, Ansari MJ, Al Naggar Y, Hassan S, Nighat S, Zehra SB, et al. Causes and Reasons of Insect Decline and the Way Forward. London: Intechopen; 2021

[98] Niemelä J, Kotze DJ, Venn S, Penev L, Stoyanov I, Spence J, et al. Carabid beetle

assemblages (Coleoptera, Carabidae) across urban-rural gradients: An international comparison. Landscape Ecology. 2002;**17**:387-401

[99] Ishitani M, Kotze DJ, Niemelä J. Changes in carabid beetle assemblages across an urban-rural gradient in Japan. Ecography. 2003;**26**(4):481-489

[100] Sadler TD, Donnelly LA. Socioscientific argumentation: The effects of content knowledge and morality. International Journal of Science Education. 2006;**28**(12):1463-1488

[101] Bonney R, Shirk JL, Phillips TB, Wiggins A, Ballard HL, Miller-Rushing AJ, et al. Next steps for citizen science. Science. 2014;**343**(6178):1436-1437

[102] Silvertown J. A new dawn for citizen science. Trends in Ecology & Evolution. 2009;**24**(9):467-471

[103] Sauermann H, Franzoni C. Crowd science user contribution patterns and their implications. Proceedings of the National Academy of Sciences. 2015;**112**(3):679-684

Chapter 7

Impact of Modern Forestry and Climate Change on Saproxylic Insect Diversity: Is Life in Dead Wood at Risk?

Sigmund Hågvar and Frode Ødegaard

Abstract

On a world basis, about one-third of forest-living insect species are saproxylic, that is, they depend directly or indirectly on dead wood. They represent many different insect groups, but the highest species numbers are found among Coleoptera, Diptera, and Hymenoptera. Many saproxylic species are red-listed due to the impact of modern forestry. Preserving their high diversity represents a big challenge in today's conservation work for insects. There is an urgent need to protect the last, deadwood-rich natural forests and to increase the amount and diversity of dead wood substrate in managed forests. We present a case study from boreal forests in Fennoscandia to illustrate how certain measures have been implemented recently to protect, or restore, habitats for saproxylic beetles. In a second case study, we describe the challenge of conserving the unique insect fauna of old, hollow oaks. Climate change can affect saproxylic insects in several ways, directly or indirectly. Summer drought could initiate large-scale forest fires. Wind felling and snow damage may increase, and other tree species could be favored. Such changes will affect the amount and diversity of dead wood substrates. Higher temperatures may favor bark beetle outbreaks and root rot in boreal forests, and the geographical distribution of species may change. The present system of protected forests in Europe is not sufficient for conserving the diversity of saproxylic insects. Stronger initiatives are needed to protect more forests, and increased considerations must be taken in forestry operations.

Keywords: beetles, case studies, certification, climate change, coniferous, conservation, dead wood, Fennoscandia, forest, forestry, habitat loss, hollow oaks, insects, saproxylic

1. Introduction

Decomposing wood contains very rich communities of insects, fungi, bacteria, and other organisms (**Figure 1**). In fact, a dead tree can harbor more species than a living one. Organisms living in dying or dead wood are called saproxylic [1]. Their ecological function in forest habitats is to recycle nutrients stored in the wood.

Figure 1.
There is plenty of life in dead wood. Left: a rich community of fungi, mosses, and lichens on a fallen stem. Right: This beetle is Cucujus cinnaberinus *(Scopoli, 1763) of family Cucujidae. It is red-listed in Norway as "Near threatened" and develops under the bark of old, newly dead aspen trees* (Populus tremula *L.). (Photo left: S. Hågvar. Photo right: Anne Sverdrup-Thygeson).*

Insects contribute to this process through a variety of feeding groups: bark-feeders, fungi-feeders, detritivores, parasitoids, predators, or commensals [2].

Saproxylic species have evolved independently within many different insect groups, but the highest numbers of species are found among Coleoptera, Diptera, and Hymenoptera. Two-thirds of the Coleoptera families and about half of the Diptera families contain at least one saproxylic species [3, 4]. Among cavity-nesting bees and wasps, one-fourth prefer cavities in dead wood [5]. Ants using dead wood cavities for nesting (**Figure 2**) support specialized guest species from various insect groups [6]. Termites are a well-known group among wood-feeders. On a world basis, about one-third of all forest insect species depend directly or indirectly on dead wood. Ulyshen and Sobotnik [7], which makes forests unique ecosystems. Norwegian studies have concluded that Mycetophilids (fungus gnats) are good indicators of forest continuity, as they usually depend on wood-living fungi, often in wet microhabitats [8].

During the former decade, two large books set focused on the rich biodiversity in dead wood, stressing the need for cooperation and conservation measures. Stokland et al. [9] produced the first book to synthesize the natural history and conservation needs of wood-inhabiting organisms in general: "Biodiversity in dead wood". That book included as diverse organisms as birds, insects, fungi, lichens, and mosses. A few years later, Ulyshen [10] highlighted the multitude and vulnerability of dead wood-depending insects, by editing the monumental book "Saproxylic Insects". In addition, there is a rich literature on special aspects around insects depending on dead wood in some parts of their life cycle. While dead wood was regarded as a wasted resource a few decades ago, there is now a broader understanding that dead wood is an important biodiversity resource—and biological hotspot.

One may ask: Why present a chapter about saproxylic insects in the present book when the topic was treated in the two last mentioned books not many years ago? Firstly, saproxylic insects are currently under pressure worldwide due to forestry activity. Secondly, interesting things are currently going on in Fennoscandian boreal forests, where forestry shows an increasing will to preserve dead wood as a habitat. This topic will be elaborated on later as a separate case study. Moreover, we will set focus on hollow trees, for instance old, hollow oaks (*Quercus* L.), which are well known for their unique insect fauna. Consciousness about this is increasing across Europe, at the same time as old trees are cut faster than they are recruited, which may take several hundred years. Interesting studies have recently been performed regarding beetles in hollow oaks. Even in Norway, in the northern part of the oak's

Figure 2.
This dead spruce tree Picea abies *(L.) H. Karst. in southern Norway harbors a colony of the Hercules ant* Camponotus herculeanus *(Linnaeus, 1758). Many holes have been made by the black woodpecker* Dryocopus martius *(Linnaeus, 1758) which preys on the ants, catching them with its long, sticky tongue. (Photo: E. Hågvar).*

distribution, an interesting beetle fauna still exists within the dark cavities of old trees, but the number of such trees is declining. Is there a sustainable future for this specialized beetle fauna in Norway, Sweden, and Europe?

For ecologists, saproxylic insects represent a treasure chest, for nature conservationists a headache, and for politicians a rather unknown topic. For forest owners, conservation measures for these insects may imply reduced income. It is necessary for ecologists, conservationists, politicians, and forest owners to come together and cooperate. The basis for such cooperation is a common ecological understanding of dead wood as an important habitat for many threatened and vulnerable insects. Before we get into the two case studies mentioned, we shall have a closer look at the ecology and vulnerability of saproxylic insects.

2. Ecology and vulnerability of saproxylic insects

During the last few decades, saproxylic insects have achieved increased attention due to the loss of natural forests in many countries (**Figure 3**). In 1999, the Council of Europe published a warning about this vulnerable group of organisms [1].

Figure 3.
If an old forest is clearcut, the habitat may need hundreds of years to restore, and many specialized insect species will probably never return. From a clearcut area in a Norwegian forest landscape dominated by spruce. (Photo: Jørn Bøhmer Olsen).

Without correct conservation measures, species may go extinct before their ecology, adaptations, and role have been revealed – or even species described. Certain species are fascinating due to their shape, color, or size. Among them are the European Eremite beetle *Osmoderma eremita* (Scopoli, 1763) which will be closer presented later.

To understand the high diversity of saproxylic insects and their problems in "modern" forests, we must understand both their evolutionary history and the ecology of natural forests. Insects living in dead wood have evolved and specialized during a very long period, from the first trees in Devon, about 400 million years ago [7]. As a functional group, saproxylic insects have a much longer evolutionary history than pollinators, which radiated together with the evolution of flowers after the extinction of Dinosaurs, about 65 million years ago [7]. The large amounts and great variation of dead wood substrates in natural forests allowed for the evolution of saproxylic species within many insect groups.

The long evolutionary history of saproxylic insects has led to a high number of specializations, both due to a high species numbers of trees, and diversity in dead wood substrates. Examples are various dimensions of dead wood, standing or lying, stage of decomposition, presence of fungal structures, as well as local moisture conditions. In certain cases, single large and old trees can harbor a variety of dead wood niches. For instance, old oak trees may have dead branches of different dimensions both on the tree and on the ground, as well as various cavities in stems and branches.

Many saproxylic insects have relationships to fungi, both as food and as habitat. Wood itself contains mostly cellulose and lignin, and very little of minerals. However, minerals may accumulate in dead wood with time, due to three processes [11]: Carbon loss during decomposition leads to mineral concentration, N-fixing procaryotes contribute to nitrogen availability, and fungi transport nutrients from the outside, for instance from soil, through their hyphae.

Relationships between wood-living fungi and insects go both ways. A literature review indicated that insects, for instance, beetles, may disperse fungal spores to new substrates by carrying them in their gut or on the body surface. For the fungus, such dispersal would be more precise than a random dispersal of spores into the air [12]. Close studies have confirmed that saproxylic beetles can function as targeted vectors for decomposer fungi [13, 14]. In field studies, aspen logs (**Figure 4**) where beetles were excluded had a different fungus flora and decomposed slower than other logs [15].

Figure 4.
*This illustration shows how a fallen stem of aspen (*Populus tremula *L.) is colonized by different species of insects, lichens, and mosses during the various phases of decomposition. The upper log has been dead for 1–2 years, the middle log for about 10 years, and the lower one for about 25 years. The following beetles are shown: On the upper log* Agrilus suvorovi *Obenberger, 1935, on the middle log* Platycerus caprea *(Degeer, 1774) and* Agathidium discoideum *Erichson, 1845, and on the lower log* Melanotus castanipes *(Paykull, 1800) and* Cychrus caraboides *(Linnaeus, 1758). Characteristic invascular organisms are shown in encircled spots. On upper log, there are two fungi:* Encoelia furfuracea *(Roth) P. Karst. (left) and* Peniophora rufa *(Fr.) Boidin (right). The middle log has a slime mold (Myxomyceta),* Fuligo septica *(L.) F. H. Wigg., 1780 to the right and two fungi to the left:* Lentaria epichnoa *(Fr.) Corner and* Clavicorona pyxidata *(Pers.) Doty. On the lower log, we see the lichen* Multiclavula mucida *(Pers.) R. H. Petersen. (Artist: Martin Holmer).*

Furthermore, the succession pattern of decomposer fungi in logs depended on which beetles had colonized the logs in the early phase [16].

Dispersal ability is an important trait for many saproxylic species [17, 18]. Dead wood is a substrate that gradually decomposes, and some species use it only for a short time. When the substrate provides suitable conditions for them. Saproxylic species must continuously find new substrates in order to persist in a viable population. They will go locally extinct if a specific habitat, as for instance hollow trees, are too far apart. Furthermore, isolated old forests have fewer saproxylic insect species [17]. In a mixed beech-spruce forest reserve in Switzerland, Schiegg [19] studied how the diversity of saproxylic Coleoptera and Diptera depended on the amount and connectivity of dead wood. Sites with high dead wood connectivity (i.e., short distance between dead wood items) were more species-rich than sites with the clumped distribution of the same amount of dead wood, suggesting that habitat fragmentation can reduce species richness even at a very local scale. A review paper by Sverdrup-Thygeson et al. [20] concluded that there is a large variation between different saproxylic taxa in their response to spatial and temporal dead wood patterns.

Management recommendations need to be supported by more studies. Such studies could follow different lines. In experimental setups, one could either use existing situations or manipulate the amount and quality of dead wood in the landscape. In either case, the scale would have to be considered: Shall we study the dispersal ability within meters, kilometers, or on a landscape level? A more general question in such studies could be: Shall we try to define measures that would benefit as many saproxylic species as possible, or should we focus more on the most threatened species, and their specific needs? We probably need both approaches but still have to give priority according to research capacity and funding possibilities.

On the species level, habitat requirements can be rather narrow, and conservation measures must be correspondingly precise. For instance, the large, endangered European longhorn beetle *Cerambyx cerdo* Linnaeus, 1758 depends on very old oak trees in rather open and sunny sites, typically in traditionally pasture landscapes. While larvae develop within the stem, adults are sap eaters, seeking cracks or wounds in the tree [21]. Long-lasting substrates, like large oaks, reduce the need for frequent dispersal. Downed logs in large dimensions that take decades to decompose, represent another "long-lasting" substrate. However, it may be difficult to decide whether a given species is limited by lack of habitat, or by a low dispersal ability.

If we shall improve life conditions for certain threatened species, we must offer them, or simulate, their ecological needs—including disturbances [22]. Certain saproxylic insects are adapted to take favor of natural disturbances like large-scale wind felling, insect attacks killing many trees, or wildfire (**Figure 5**). Some Coleoptera and Heteroptera are equipped with special sensors that can detect heat or smoke from wildfires from a long distance [23–27].

Continuity is a key concept. It means the continuous presence of a certain habitat, substrate, tree type, climate, or even a kind of disturbance, over time. If continuity is broken, as clearly happens for instance after clear-felling of an old forest, only species with a sufficient dispersal ability to find their specific habitat somewhere in the landscape will survive. Species with limited dispersal ability are much more threatened by fragmentation or the break of forest continuity. However, for species favored by burning, a good dispersal ability is not sufficient if burning occurs very infrequent or spread geographically.

In a mature Norwegian spruce forest, obligate saproxylic beetles caught by window traps were related to the following variables: decaying wood, wood-inhabiting

Figure 5.
Some specialized insects, for example, beetles and flat bugs, benefit from periodic forest fires. They probably feed on certain fungi that decompose the burned wood. Controlled burning of forest patches can be necessary to conserve these insects. (Photo: Jørn Bøhmer Olsen).

fungi, the level of disturbance, landscape ecology, and vegetational structure. Level of disturbance was defined by lack of continuity for tree cover and dead wood, based on data on former thinning or clearcutting, and fungal indicators. Catches were only weakly related to the dead wood amount in the nearest 40 x 40 m but several strong relationships were found at one km^2 and four km^2 levels [28]. The following attributes increased the species richness of many groups, and often increased the abundance as well: Diversity of dead tree parts, number of dead trees of large diameter, and number of polypore fungi species. A negative factor was former extensive cutting.

Lachat and Müller [29] highlighted the following six characteristics of primary forests that were important for saproxylic insects:

1. Absence of habitat fragmentation (sufficiently large area).

2. Continuity of dead wood (by continuous production and availability)

3. Natural disturbance regimes (for instance a certain natural fire frequency).

4. Presence of dead wood in sufficient amounts, and in various types, dimensions, and decomposition phases.

5. Tree species composition (some insects need specific tree species).

6. Habitat trees (for instance hollow oaks).

Summing up, protection of natural forests, preferably on a landscape scale, is obviously the best conservation measure. This is, however, rarely possible. Instead, our challenge may be to identify and protect the last, small remnants of natural forest.

Here, relict populations of the most sensitive species may have survived. Furthermore, we could lengthen rotations to allow harvested forest to get older. Controlled burning may be done to simulate a natural disturbance. This will not only produce burned wood that some specialists need, but burned ground may favor the establishment of short-lived pioneer deciduous trees. After a few decades, dead wood from these trees may offer substrate for several saproxylic insects, including species that breed in, or feed on, fruiting bodies of polypore fungi. Retaining dead wood and high stumps, and avoiding dead wood harvesting, would obviously be favorable.

Climate change may affect saproxylic insects in many ways. This topic is treated in Section 5.

3. Case: Spotlight on saproxylic beetles in Fennoscandian boreal forests. Threats, measures, and possibilities

3.1 Geography, forest types, and natural processes

To illustrate today's challenges and efforts for conserving saproxylic insects, we have chosen a case about beetles in Fennoscandian forests. During the last decades, much knowledge has accumulated about the ecology and threats of saproxylic beetles in Norway, Sweden, and Finland. Furthermore, there is an ongoing discussion on which measures should be taken in forestry operations to improve living conditions for threatened species.

A common feature of Norway, Sweden, and Finland is their considerable areas of boreal coniferous forests, representing the western outpost of the Eurasian taiga. In all three countries, major parts of these forests have been clearcut at least once. This means that their "continuity" has been broken. During thousands of years, a "continuous" presence and production of dead wood substrates secured life conditions for a high diversity of decomposing insects, fungi, and bacteria, including species with a limited dispersal ability. Dead wood even serves as a substrate for specialized mosses and lichens [9]. Clearcutting of these old natural forests reduced or removed life conditions for saproxylic species that depended on large, old trees or the continuous production of various dead wood substrates. Only species with high dispersal ability, tolerance for open areas, or ability to use smaller dimensions of dead wood, could survive in the new type of forest landscape.

Several studies have documented a tight connection between saproxylic fungi and beetles (**Figure 6**). For instance, fruiting bodies of two common polypore species, *Fomitiosis pinicola* (Swartz:Fr.) Karst. and *Fomes fomentarius* (L.:Fr.) Kickx harbor quite different beetle communities, both during living and dead stages [31–33]. At least some beetles were shown to use the odor from the fruiting bodies to navigate toward the correct fungus during flight [34].

Lowland, productive forests are generally the most species-rich, both regarding saproxylic beetles and other organism groups. This is partly due to a natural mix with various deciduous trees, like birch (*Betula* L.), aspen (*Populus tremula*), goat willow (*Salix caprea* L.), rowan (*Sorbus aucuparia* L.), and gray alder (*Alnus incana* (L.) Moench). In all three countries, nearly all lowland forests have been subject to harvesting.

In Sweden, boreal coniferous forests are restricted to the northern and medium parts. Southern Sweden has a warmer climate than Norway and Finland and contains considerable areas of warm-demanding deciduous trees, such as oak (*Quercus* sp.),

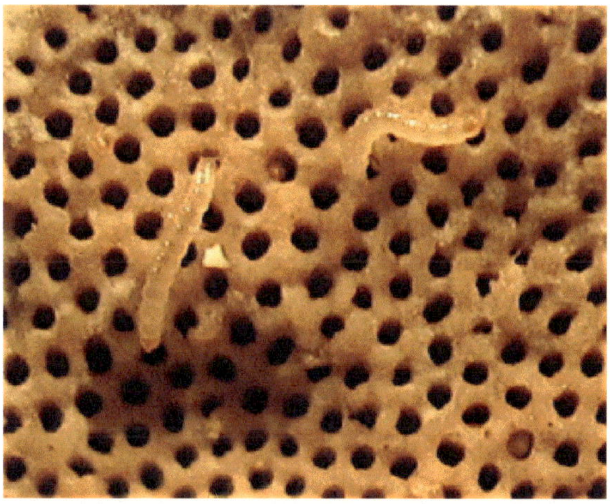

Figure 6.
Larvae of the staphylinid beetle Gyrophaena boleti *(Linnaeus, 1758) live inside the pores of the fungus* Fomitopsis pinicola. *Here, two of them are shortly on the surface while they change pores [30]. (Photo: S. Hågvar).*

maple (*Acer* L.) ash (*Fraxinus excelsior* L.), and beech (*Fagus sylvatica* L.). In certain places, single hollow oaks on private lands still exist due to long-term protection by landowners. Hollow oaks are famous for containing a specialized insect community. While most of this chapter about saproxylic Fennoscandian beetles will concentrate upon boreal coniferous forests, the insect fauna of hollow oaks will be closely discussed at the end of the chapter.

3.2 Species numbers, threats and red lists

A review article by Siitonen [35] described well the importance of coarse woody debris (CWD) for the survival of saproxylic organisms in Fennoscandian boreal forests. It was estimated that at least 4000–5000 organisms depended on dead wood habitats in Finland, accounting for 20–25% of all forest-dwelling species. The three most species-rich groups were beetles with around 800 saproxylic species, and Diptera and Hymenoptera with 500–1000 saproxylic species each. Reasonably, areas with large amounts of CWD supported the highest species numbers. Trunks of large diameter were especially valuable, preferably in medium and advanced stages of decay. Protection of the last old-growth patches in the landscape were of high value for saproxylic species, not only because of large amounts and high diversity of dead wood but also due to a historic continuity of dead wood substrates. Compared to natural forests, the average amount of CWD in managed forests had probably been reduced by 90–98% in Finland [35]. By using general species-area relationships, this reduction in substrate availability indicates that more than half of the original saproxylic species will disappear in the long-term [35].

Looking at the red lists from the three countries, we can extract the following information (e. g. [36]): About one fifth of the beetle species in these countries are red-listed. For Norway, there are 826 red-listed beetles, of which 45% are forest-living. The great majority of the latter are dependent upon dead wood [36]. Modern forestry is the greatest threat because it removes the habitat of these species. Even some of the oldest forests are still subject to forestry. Too few forests are protected and the oldest ones should be the first to be protected. Many species are under reduction in distribution and abundance. For instance, in Norway, the area of productive, "natural" forest that has never been clearcut, was reduced by 28% between 1990 and 2016 [37]. Altogether, two-thirds of the total forested area in Norway have now been subject to clearcutting. In Sweden, 60% of the productive forests were clearcut between 1950 and 2010 [38]. In all three countries, the amount of dead wood is low even in the forests that have never been clearcut, due to selective cutting during long times. This is why Fennoscandia is a good place to test dead wood restoration programs aimed to enhance saproxylic diversity. Studies from Norway show that the amount of dead wood has increased during the period 2002–2017, since harvesting volume has been lower than natural production [39]. That is a positive development but the amount is still only 15–20% of that in natural, old forests. Furthermore, dead wood in natural forests is of a different character, with larger dimensions, other fungal species, and for instance hollow trees of very high age.

If we look at those beetle species on the red lists that, in strict terms, are called "threatened", we find 452 such beetles from Norway, 391 from Sweden, and 288 from Finland. Several of these are common to two or three countries. "Threatened" species is here a combination of the three highest categories in The International Union for Conservation of Nature (IUCN): critically threatened (CR), endangered (EN), and vulnerable (VU).

3.3 Protected areas

In 2016, the Norwegian Parliament decided that 10 percent of the forest area shall be protected. This was celebrated by nature conservation organizations as a large victory. However, this aim was neither followed up by a time plan nor a financial guarantee. During the following eight years, the protected area has increased only slowly, from 3 to 5 percent. In 2024, the conservation process is still slow, mainly due to limited financing by government for compensation to forest owners. Many forest owners are, in fact, interested in "selling" their forest for conservation purposes, but the forests offered are often among the less productive and less valuable for protection (**Figure 7**). That is a real problem. The forest owner organization has demanded the privilege of proposing forests for protection and has threatened to boycott all further forest protection work if they lose that privilege. While voluntary forest protection proceeds slowly, parts of the oldest, most valuable forests are cut – partly supported by official money.

In Sweden, 9 percent of the forest area was under protection at the end of 2021. Most of this is situated at relatively high altitudes along the long, western mountain ridge. Only small amounts of the productive forests have been protected. In Finland, 13 percent of the forest area had been protected at the start of 2022. Three quarters of these forests are situated in northern parts of Finland, where species richness is lowest. There is a need in all three countries to protect more of lowland, productive forest. To achieve this, political initiative is necessary.

3.4 Measures in forestry (boreal, coniferous forest)

3.4.1 Certification of forestry operations

Today, forest operations in Norway, Sweden, and Finland are certified on a voluntary basis according to certain standards to favor biodiversity. Certified timber is expected to achieve a higher price. The standards vary and are a result of negotiations

Figure 7.
Insects depending on old trees and dead wood need large, protected forests to survive, such as in this coniferous forest reserve of about 20 km² (Skirvedalen nature reserve, Southern Norway). (Photo: S. Hågvar).

between forest stakeholders and environmental organizations. Several systems exist, of which the most renowned is that of The Forest Stewardship Council (FSC). In Fennoscandia, the importance of dead wood is expressed by retaining old trees, snags, windthrows, and other dead wood elements like coarse woody debris. Leaving high stumps is even practiced. There are also general biodiversity-friendly practices, as retaining forest strips along rivers, lakes and bogs, and to set aside patches especially suited for survival of red-listed species, so-called key habitats. These general measures often contribute to the amount and diversity of dead wood habitats.

There is a history behind this certification. From the 1980s on, Fennoscandian entomologists pointed out the danger of losing specialized insects depending on old-growth forests, which declined in area in all three countries. Two international agreements fueled further awareness: The 1987 report "Our common future", launching the concept of sustainable development, and the Convention on biodiversity agreed upon in Rio in 1992. These sparked various research projects on biodiversity in Fennoscandian boreal forests. In 1997, ecologists came together in Uppsala, Sweden, to a conference about "Biodiversity in managed forests". A central point was the importance of coarse woody debris (CWD) for maintaining biodiversity. Two years later, a "Nordic symposium on the ecology of coarse woody debris in boreal forests" was held, attracting 70 persons from 11 countries. The presentations were later collected under the title "Ecology of woody debris in boreal forests", today a classic publication [40]. This formed a basis not only for further research but also for implementing new scientific knowledge into Nordic forestry practices. Since then, we have seen a growing awareness and responsibility in Fennoscandian forestry for preserving biodiversity, including saproxylic insects. However, there are still improvements to be made. Here, we shall discuss some topics which are under debate at present (2024).

3.4.2 Certification measures under debate

In Norway, most forest owners are certified according to "Programme for the Endorsement of Forest Certification", the so-called PEFC forest standard (cdn.pefc.org), which are revised every fifth year. According to forest owner organizations, this standard guarantees that the timber has been harvested in a sustainable way, including the conservation of biological diversity. Let us have a closer look at the so-called key habitats, that shall remain intact since they are assumed to be important for red-listed species. Several of the parameters used to identify key habitats have relevance for saproxylic insects: Standing dead wood, lying dead wood, patches with old deciduous trees, old trees in general, hollow trees, and burned forest areas. For saproxylic insects, the way key habitats are identified is of importance. Recently, Norwegian researchers and nature conservation organizations have strongly criticized the praxis around key habitats, including how they are registered (the so-called MiS method):

- Registrations of key habitats are not made by biologists. Only indirect parameters like the presence of dead wood or old trees are used, and species are not identified. Detailed studies by skilled biologists have shown that nonbiologists detect only one-seventh of biologically important environments [41].

- After registration, the forest owner is free to decide which key habitats should be conserved or not, and this selection is not made in an open process.

- Although more than 70,000 key habitats have been set aside in Norway, their mean size is only 10.7 da. Such small patches surrounded by open clearcuts may dry partly out, be subject to wind felling, and may contain small, nonviable populations of the species they are meant to save. They may also be situated too far from each other to exchange genetic material for the actual species.

- Bureau Veritas shall control that key habitats are properly identified and conserved, but such control is only sporadically.

- Corridors connecting key habitats are lacking. Near Oslo, the local section of The Norwegian Society for the Conservation of Norway has proposed to establish a such corridor by allowing younger forest between old-growth key habitats to develop freely toward old and dead wood-rich sites. The project is named "Continuous wilderness" [42]. Part of this forest is owned by the municipality of Oslo, and the politicians there are in principle positive to the corridor proposal. Private forest in the actual area makes it, however, difficult to achieve the full aim.

Furthermore, in Norway, it is difficult for nature conservation bodies to evaluate or stop a planned cutting:

- The time period available to comment upon a planned cutting, is too short for checking the site for possible-red-listed species.

- Registration of key habitats, the creating of certification rules, and final permission to start cutting are all processes that the "forest community" controls. During the latest certification changes, several nature conservation bodies, therefore, denied to participate.

- The Norwegian Society for the Conservation of Nature has argued that prior to any cutting activity, the actual forest patch should always be investigated by biologists for possible-red-listed species.

In Finland, a study has been performed on the ability of key habitats to conserve the diversity of wood-decaying polypore fungi [43]. Several saproxylic beetles use polypore fruiting bodies for larval development or for feeding. On average, the key habitats hosted more polypore species than control sites. However, since very few red-listed fungi were documented, key habitats seemed to be of little help in the conservation of threatened polypore fungi.

From this discussion, several research questions arise: Should threatened species be registered independently of key habitats? Do the red-listed species survive in their small key habitats? What is the dispersal ability of different species, and do they need ecological corridors? Is there a lower limit for the amount (and quality) of dead wood that is necessary for certain species to survive? Is regularly transport of dead wood into certain sites a good idea to simulate continuity? How to ensure more frequent control of practicing the certification rules? How to achieve a more objective evaluation of environmental values prior to cutting, and give environmental bodies a chance to evaluate the plans and protest to them?

Another question is about the effect of retaining trees on clearcut flats.

According to the Norwegian standard, at least ten storm-resistant trees shall be left per ha and allowed to grow and die naturally. In this respect, it may be of great significance which tree species you leave on the clearcut. Nordic coniferous forests contain several deciduous trees, both as a natural element during early succession, and as a variation in mature forest. A study by Sverdrup-Thygeson and Ims [44] illustrated the value of retaining aspens on clearcut. Being heated by the sun, dead stems could be habit for several red-listed beetles. Standing snags were richer on saproxylic beetles than in logs.

Experimental evidence on the biodiversity impacts of retention forestry, prescribed burning, and deadwood manipulation in Fennoscandia was summed up by Koivula and Vanha-Majamaa [45]. Dead wood may be allowed to accumulate over time, and be supplemented by girdling trees to produce snags that could be left after cutting. By good planning of felling operations, the continuum of dead wood can to a certain degree be maintained and help several saproxylic species to be "life-boated" over the regeneration phase. Such awareness and practice depend on educated foresters. Even adding extra amounts of dead wood by leaving some of the cut stems is a possibility, although the species composition in these logs will differ from that of naturally died trees. Certain beetles that are favored by relatively warm dead wood may profilerate on clearcut areas. However, many characteristic old-forest species demanding continuity in a moist climate, and perhaps dependent on special types of dead wood, including very old trees (**Figure 8**), cannot be compensated for by various types of retention.

Restoration of managed stands is to a certain degree possible over time. This should preferably be done close to protected areas, to extend the living area of sensitive species [35].

Figure 8.
An old, dying spruce tree in southern Norway. Such trees, and the dead wood they produce, are important for many species of saproxylic insects. (Photo: S. Hågvar).

The dialog which has been developed between researchers, forest owners, and forest authorities in Fennoscandia can be a good model for other countries, including certification rules for sustainable forestry. Several open questions remain, however, regarding good measures for saproxylic insects. For instance, forest managers ask «how many snags do you need per ha or per 100 ha? How much dead wood volume is needed? What is a suitable frequency of fire?" Sufficiently large area to burn can be 1 ha for a certain species, 10 ha for another and > 1000 ha for a third one. There is a trade-off between the optimal measures for biodiversity on one hand, and the willingness among forest owners to sacrifice a certain loss of income.

3.4.3 How to conserve fire-favored saproxylic insects?

Among Coleoptera, Diptera, Heteroptera, and Lepidoptera there are several fire-favored (pyrophilous) species in Fennoscandian boreal forest [46, 47]. Most of these are saproxylic, using burned wood for their development during the two first years after burning. Wikars [47] concluded that the most specialized species are either feeding on certain ascomycete fungi that are favored by burning, or they may be phloem-feeders or predators favored by some habitat characteristics in newly burned forests other than burned wood. His study showed that pyrophilous insect species were almost exclusively confined to burned forests. However, surprisingly, such species occurred in both burned and unburned logs on burned sites. Hjältén et al. [48] suggested that certain fire-adapted species must be able to maintain viable populations in the unburned forest matrix if it is of sufficient quality. In other words, maybe the sum of all conservation efforts made in the whole landscape, and not fire restoration alone, may support low populations of what we call pyrophilous insects. Further studies may help to focus on the most relevant research questions and the best future burning strategies.

In Sweden, controlled forest burning has been performed yearly since a plan for conserving fire-depending insects was developed [49]. This plan, for the period 2006–2010, was described like this:

> *This action plan is targeted at favoring red-listed fire-dependent insects in the majority of the boreal forest in Sweden. The biology and distribution of four species of bugs (Hemiptera: Aradidae;* Aradus angularis *J. Sahlberg, 1886,* A. aterrimus *Fieber, 1864,* A. laeviusculus *Reuter, 1875 and* A. signaticornis *R. F. Sahlberg, 1848), five beetles (Coleoptera:* Biphyllidae, Biphyllus lunatus *(Fabricius, 1787); Bostrychidae,* Stephanopachys linearis *(Kugelann, 1792) and* S. substriatus *(Paykull, 1800); Latrididae,* Corticaria planula *Fall, 1899 and Cerambycidae,* Acmaeops marginata *Mroczkowski, 1986) and one fly (Diptera:* Empididae, Hormopeza obliterata *Zetterstedt, 1838) are described. Four species are Nature 2000-species (*A. angularis, S. linearis S. substriatus, *and* C. planula*). The species depend on fire-damaged trees or, in one case (the fly), on deeply burned soil.*

An evaluation in 2024 by Naturvårdsverket of this project gave the following conclusion [50]. During the period 2016–2019, controlled forest burnings, from 4 to 50 ha in area, were performed all over Sweden, mainly in a number of protected areas. In several sites, positive effects of burning were confirmed on fire-depending insects. There is, however, a discussion going on regarding further strategy. Such a project is demanding to organize, people must be educated to perform burning safely, money are needed, burning should perhaps be done every second year, and maybe each burning should be designed to preserve one or certain species that may be very local or have special demands to substrate, for instance large diameter of wood. Anyhow, Sweden has for some decades taken the threat about fire-favored insects seriously. Their measures may guide and inspire other countries in this field.

In Finland, controlled burning has in fact been a part of ordinary forestry since the 1950's. The purpose was originally to prepare the soil well for a new generation of trees, but this practice has also favored pyrophilous insects. Nowadays, controlled burning is more related to nature conservation than to forest management. In Finland, PEFC certification requires burning if the certified forest covers an area of more than 200,000 hectars. The owner of an FSC-certified forest must perform controlled burning if the property is larger than 10,000 hectars. Positive effects on fire-dependent insects have been recorded. The first flying insects might arrive already by nightfall after a controlled burning, for instance, the jewel beetle *Melanophila acuminata* (DeGeer, 1774) [51]. However, in 2018, it was claimed that Finland has a problem with too few forest fires to protect biodiversity. Furthermore, too few burned-down areas are protected [52]. Today, EU funding enables restoration burning in Finland. It has been documented that plenty of species dependent on deadwood always arrive on fire sites from the nearest forests, especially from old adjacent forests, where they inhabit naturally created deadwood. Clearly, burned forests have a general positive effect on populations of many deadwood-depending insects, not only the strictly pyrophilous species [53].

In Norway, no systematic forest burning for the purpose of biodiversity is practiced. However, studies performed on naturally burned sites have confirmed the colonization of several pyrophilous insect species [54–58].

4. Case: Spotlight on hollow oaks: A special habitat with a vulnerable insect fauna

Large hollows in oaks (**Figure 9**), or cavities in other long-lived deciduous trees, may contain a specialized and species-rich assemblage of saproxylic insects, especially Coleoptera and Diptera. A good overview of this special habitat, and its inhabitants, was given by Mico [59]. Such old, hollow oaks are now generally rare throughout Europe and especially their threatened beetle fauna has achieved considerable attention. Some of these beetles are big, spectacular, and famous, such as the European stag beetle, *Lucanus cervus* (Linnaeus, 1758), and the eremite beetle, *Osmoderma eremita.* Certain landscapes in southern Sweden, having a warmer climate than Norway and Finland, still have sites containing many hollow oaks. The warmest, southernmost parts of Norway also contain some old and hollow oaks, supporting a limited fraction of this specialized beetle fauna [60–62].

In Norway, hollow oaks have achieved status as a "selected habitat type" and are given general protection under the terms of §52 in the Nature Diversity Act and associated regulations [63]. Behind this decision lies a number of detailed studies on the unique beetle fauna in Norwegian hollow oaks. Today's situation for beetles in hollow oaks is a result of the long-lasting effects of logging [60]. According to Hatlevoll et al. [64], the number of hollow oaks in southernmost Norway is declining, being reduced

Figure 9.
Such old, hollow oak trees contain very specialized insect communities. The picture is from "Berge landscape protected area" in Kvam, Southern Norway. (Photo: Harald Bratli).

by 5% during a period of 3–7 years. For the long-term protection of the unique insect fauna in Norwegian hollow oaks, remaining veteran trees should be better monitored and protected. Furthermore, we must secure a continuity of old oaks over large areas, in different environments, and in sufficient density. This means that young oaks must be allowed to age and develop hollows. That is a very slow process, often initiated by certain fungi (**Figure 10**). Furthermore, geographical and landscape parameters are significant. The eastern and western regions of southern Norway host different community structures, with fewer species, also of red-listed ones, in the western region [65]. Even the surroundings of each hollow oak matter. Comparing beetle communities within oaks standing in open landscapes or in forests, Sverdrup-Thygeson et al. [66] concluded that community structure in the two situations only partly overlaps. Therefore, both single oaks in parks, in cultural landscapes, and within forests, should be protected. In more detail, Wetherbee et al. [67] demonstrated that beetle functional diversity could be related to specific parameters. Predator species richness was related to the regrowth of shrubs. Decomposer species richness responded to tree vitality, and functional diversity was related to both tree circumference and habitat connectivity.

In Europe, *Osmoderma eremita* (**Figures 11** and **12**) has become a "flagship species" for the conservation of hollow trees and the threatened community that these contain. According to Annex II of the EU Habitats Directive, 1992, *O. eremita* is protected by European law. The species also functions as an "umbrella species" since measures to protect the eremite beetle may preserve the stable hollow-tree habitat needed for a number of other threatened insects.

Figure 10.
This yellow fungus, called chicken of the woods (Laetiporus sulphureus *(Bull.) Murrill), typically lives in decaying oak stems, contributing to the gradual formation of hollows. (Photo from Southern Norway, by Anne Sverdrup-Thygeson).*

Figure 11.
*The eremite beetle (*Osmoderma eremita*) rarely leaves the cavity in which it developed. Generations follow generations within the same hollow oak tree. (Photo: Arnstein Staverløkk).*

Since hollow oaks may persist for hundreds of years, the actual species live in a very stable habitat and have little need for dispersal. *O. eremita* is both dispersal-limited and habitat limited. It is a short-distance disperser, usually remaining within the same hollow stem [68]. However, sooner or later, the species must disperse to another hollow oak. A successful long-term conservation presupposes both a certain density of such old trees in the landscape, as well as a continuous production of younger trees that are allowed to become old. If such oaks stand close enough to allow insect dispersal between them, forming a metapopulation dynamic, certain very rare species can still exist and thrive in the old tree cavities. Fortunately, it is possible to help hole-living species to survive by offering surrogate tree cavities. According to Hilszczanski and Jaworski [69], the red-listed species *Osmoderma barnabita* Motschulsky, 1845, and probably other species typical for hole oaks, can breed temporarily in "bird-boxes" with artificial substrate.

Whether the EU directive is sufficient to halt the decline of *O. eremita* and its allies in hollow trees (**Figure 13**), including the remaining population in southern Sweden, remains to be seen. Extinction is often a gradual process. If new hollow oaks are not continuously recruited, the forest landscape will sooner or later be unsuitable for this specialized insect community. According to Hylander and Ehrlén [70], species that are still present but may be doomed to extinction, are paying an "extinction debt".

Figure 12.
*An entomologist is looking for a possible occurrence of the eremite beetle (*Osmoderma eremita*) in a hollow ash tree (*Fraxinus excelsior*) in Southern Norway. (Photo: Arnstein Staverløkk).*

Figure 13.
Left picture: Elater ferrugineus *Linnaeus, 1758, a rare click beetle (Elateridae) living in hollow trees, mainly oak. Photo: Arnstein Staverløkk. Right picture:* Protaetia marmorata *(Fabricius, 1792) is also a hollow-tree specialist, developing in cavities of oak or ash. Photo: Oddvar Hanssen.*

5. Climate change and saproxylic insects

Climate change may influence Fennoscandian saproxylic beetles in several ways: by gradual temperature rise, pushing some species out of their tolerance limit, by changes in the types and diversity of dead wood, and by changes in disturbance regimes (fire, wind felling and large insect attacks). In a warmer climate, the common spruce bark beetle *Ips typographus* (C. Linnaeus, 1758) may be able to fulfill two generations per year. That would lead to higher populations and make the species a more problematic forest pest [71]. Furthermore, there can be changes in the operation practices during tree harvesting. New tree species may dominate, and the need for timber may increase, during such types of changes, the amount and diversity of dead wood, and the ability to disperse and localize a specific dead wood substrate, will be decisive for each saproxylic species.

Norway spruce (*Picea abies* (L.) H.Karst), which dominates most forest landscapes in Fennoscandia, is adapted to cold winters. Winter dormancy starts with a rest stage, which needs a certain amount of chilling to be completed and finally broken [72]. Spruce stands in southern parts of Norway, Sweden, and Finland may gradually be substituted by deciduous trees. That implies a change in types of dead wood, and perhaps dimensions and microclimate. The many bark beetles and their followers bound to spruce might have to move northwards. New groups of saproxylic beetles, including those living in dead wood of warm-loving trees, or in hollow, large deciduous trees, may also shift their distribution northwards. Such forced migration may represent a stress factor for each species. On the other hand, increased temperature could also be an opportunity for species living in deciduous trees to increase their distribution.

In a literature review, Venäläinen et al. [73] listed the following risks to Finnish forestry and boreal forests, induced by climate change. In northern Finland, the forest growth rate was estimated to increase while growing conditions in the south may become suboptimal. The risk of snow damage would probably increase in the north but southern forests would become more vulnerable for wind damage. Summer drought could initiate large-scale forest fires. A generally warmer climate may favor both bark beetle outbreaks and increased root rot in coniferous forests. A combination of large-scale wind damage and a bark beetle outbreak could have detrimental effects on the forest landscape. For saproxylic insects, a massive wind felling could create a boost in several species but would afterward represent a break in the continuity of the production of dead wood.

Della Rocca et al. [74] modeled the effect of climate change on 56 endangered Italian saproxylic beetles. They concluded that hotspots of rare species will shift over time, representing a threat factor. Furthermore, their study revealed that existing protected areas in Italy would be inadequate for assuring the conservation of these rare, saproxylic beetles under current and future climate conditions.

Likewise, a study from Romania documented that saproxylic beetles are facing problems due to future climate change. Focusing on five species, they concluded that they may lose over 80% of their suitable habitat, and their distribution would be restricted to higher elevations [75]. The present Natura 2000 network of protected areas is inadequate for the conservation of these species and the authors advocate for the expansion of the Natura 2000 sites. They refer to the goals of the EU Biodiversity Strategy 2030, which aims at protecting at least 30% of the EU's territory.

Clearly, plans for securing saproxylic insects must include additional protected forests, both to save local populations of threatened species and to allow species to adjust their distribution under climate change. Since a warmer climate may trigger a migration toward cooler areas, the establishment of protected reserves should include the northern part, or the highest altitude, of a species distribution area. In a long-term climate perspective, forestry will be challenged continuously to adjust forestry operations to combine harvesting with biodiversity conservation. Realization of such aims presupposes fruitful cooperation between biologists, forest owners, forestry practice, forest administrations, and politicians. Biologists and nature conservation organizations have a special responsibility to argue for such cooperation. Increased awareness among people would be necessary to conserve "the small creatures that run the world". If political leaders do not lead, all kinds of pressure and information "from below" will be needed.

Author details

Sigmund Hågvar[1*] and Frode Ødegaard[2]

1 Faculty of Environmental Sciences and Natural Resource Management, Norwegian University of Life Sciences, Ås, Norway

2 Department of Natural History, Norwegian University of Science and Technology, NTNU, Trondheim, Norway

*Address all correspondence to: sigmund.hagvar@nmbu.no

References

[1] Speight MCD. Saproxylic Invertebrates and Their Conservation. Strasbourg: Council of Europe; 1989

[2] Grove SJ. Saproxylic insect ecology and the sustainable management of forests. Annual Review of Ecology and Systematics. 2002;**33**:1-23

[3] Gimmel ML, Ferro ML. General overview of saproxylic Coleoptera. In: Ulyshen M, editor. Saproxylic Insects. Zoological Monographs. Vol. 1. Cham: Springer; 2018. pp. 51-128

[4] Ulyshen M, editor. Saproxylic Insects. Zoological Monographs. Vol. 1. Cham: Springer; 2018a. DOI: 10.1007/978-3-319-75937-1_17

[5] Bogusch P, Horák J. Saproxylic bees and wasps. In: Ulyshen M, editor. Saproxylic Insects. Zoological Monographs. Vol. 1. Cham: Springer; 2018. pp. 217-235

[6] King JR, Warren RJ II, Maynard DS, Bradford MA. Ants: Ecology and impacts in dead wood. In: Ulyshen M, editor. Saproxylic Insects. Zoological Monographs. Vol. 1. Cham: Springer; 2018. pp. 237-262

[7] Ulyshen MD, Sobotnik J. An introduction to the diversity, ecology, and conservation of saproxylic insects. In: Ulyshen M, editor. Saproxylic Insects. Zoological Monographs. Vol. 1. Cham: Springer; 2018. pp. 1-47

[8] Økland B. Unlogged forests: Important sites for preserving the diversity of mycetophilids (Diptera: Sciaroidea). Biological Conservation. 1996;**76**(3):297-310

[9] Stokland JN, Siitonen J, Jonsson BG, editors. Biodiversity in Dead Wood. Cambridge: Cambridge University Press; 2012. DOI: 10.1017/CBO9781139025843

[10] Ulyshen MD. Saproxylic Diptera. In: Ulyshen M, editor. Saproxylic Insects. Zoological Monographs. Vol. 1. Cham: Springer; 2018b. pp. 167-192

[11] Filipiak M. Nutrient dynamics in decomposing dead wood in the context of wood eater requirements: The ecological stoichiometry of saproxylophagous insects. In: Ulyshen M, editor. Saproxylic Insects. Zoological Monographs. Vol. 1. Cham: Springer; 2018. pp. 429-469

[12] Birkemoe T, Jacobsen RM, Sverdrup-Thygeson A, Biedermann PHW. Insect-fungus interactions in dead wood systems. In: Ulyshen M, editor. Saproxylic Insects. Zoological Monographs. Vol. 1. Cham: Springer; 2018. pp. 377-427

[13] Jacobsen RM, Kauserud H, Sverdrup-Thygeson A, Bjorbækmo MM, Birkemoe T. Wood-inhabiting insects can function as targeted vectors for decomposer fungi. Fungal Ecology. 2017;**29**:76-84

[14] Jacobsen RM, Sverdrup-Thygeson A, Kauserud H, Birkemoe T. Revealing hidden insect–fungus interactions; moderately specialized, modular and anti- nested detritivore networks. Proceedings of the Royal Society B. 2018a;**285**:20172833. DOI: 10.1098/rspb.2017.2833

[15] Jacobsen RM, Sverdrup-Thygeson A, Kauserud H, Mundra S, Birkemoe T. Exclusion of invertebrates influences saprotrophic fungal community and wood decay rate in an experimental field study. Functional Ecology. 2018b;**32**(11):2571-2582. DOI: 10.1111/1365-2435.13196

[16] Jacobsen RM, Birkemoe T, Sverdrup-Thygeson A. Priority effects of early successional insects influence late successional fungi in dead wood. Ecology and Evolution. 2015;**5**(21):4896-4905

[17] Feldhaar H, Schauer B. Dispersal of saproxylic insects. In: Ulyshen M, editor. Saproxylic Insects. Zoological Monographs. Vol. 1. Cham: Springer; 2018. pp. 515-546

[18] Ranius T. Measuring the dispersal of saproxylic insects: A key characteristic for their conservation. Population Ecology. 2006;**48**:177-188

[19] Schiegg K. Effects of dead wood volume and connectivity on saproxylic insect species diversity. Écoscience. 2000;**7**(3):290-298. DOI: 10.1080/11956860.2000.11682598

[20] Sverdrup-Thygeson A, Gustafsson L, Kouki J. Spatial and temporal scales relevant for conservation of dead-wood associated species: Current status and perspectives. Biodiversity and Conservation. 2014;**23**:513-535

[21] Buse J, Schröder B, Assmann T. Modelling habitat and spatial distribution of an endangered longhorn beetle—A case study for saproxylic insect conservation. Biological Conservation. 2007;**137**(3):372-381

[22] Esseen P-A, Ehnström B, Ericson L, Sjöberg K. Boreal forests. Ecological Bulletins. 1997;**46**:16-47

[23] Bell AJ. Like moths to a flame: A review of what we know about pyrophilic insects. Forest Ecology and Management. 2023;**528**:120629. DOI: 10.1016/j.foreco.2022.120629

[24] Klocke D, Schmitz A, Soltner H, Bousack H, Schmitz H. Infrared receptors in pyrophilous ("fire loving") insects as model for new un-cooled infrared sensors. Beilstein Journal of Nanotechnology. 2011;**2**:186-197

[25] Saint-Germain M, Drapeau P, Buddle CM. Persistence of pyrophilous insects in fire-driven boreal forests: Population dynamics in burned and unburned habitats. Diversity and Distributions. 2008;**14**(4):713-720. DOI: 10.1111/j.1472-4642.2007.00452.x

[26] Schmitz A, Schätzel H, Schmitz H. Distribution and functional morphology of photomechanic infrared sensilla in flat bugs of the genus Aradus (Heteroptera, Aradidae). Arthropod Structure & Development. 2010;**39**(1):17-25. DOI: 10.1016/j.asd.2009.10.007

[27] Schmitz H, Soltner H, Bousack H. Insect infrared sensors. In: Bhushan B, editor. Encyclopedia of Nanotechnology. Dordrecht: Springer; 2012. DOI: 10.1007/978-90-481-9751-4_263

[28] Økland B, Bakke A, Hågvar S, et al. What factors influence the diversity of saproxylic beetles? A multiscaled study from a spruce forest in southern Norway. Biodiversity and Conservation. 1996;**5**:75-100. DOI: 10.1007/BF00056293

[29] Lachat T, Müller J. Importance of primary forests for the conservation of saproxylic insects. In: Ulyshen M, editor. Saproxylic Insects. Zoological Monographs. Vol. 1. Cham: Springer; 2018. pp. 581-605

[30] Hågvar S. Contribution to the ecology of *Gyrophaena boleti* (Linnaeus, 1758) (Coleoptera, Staphylinidae) breeding in the pore layer of the fungus *Fomitopsis pinicola* (Fr.) karst. Norwegian Journal Of Entomology. 2018;**65**:108-114

[31] Hâgvar S. Saproxylic beetles visiting living sporocarps of *Fomitopsis pinicola* and *Fomes fomentarius*. Norwegian Journal Of Entomology. 1999;**46**:25-32

[32] Hågvar S, Steen R. Succession of beetles (genus *cis*) and oribatid mites (genus *Carabodes*) in dead sporocarps of the red-banded polypore fungus *Fomitopsis pinicola*. Scandinavian Journal of Forest Research. 2013;**28**(5):436-444. DOI: 10.1080/02827581.2012.755562

[33] Jonsell M, Nordlander G, Ehnström B. Substrate associations of insects breeding in fruiting bodies of wood-decaying fungi. Ecological Bulletins. 2001;**49**:173-194

[34] Fäldt J, Jonsell M, Nordlander G, et al. Volatiles of bracket fungi *Fomitopsis pinicola* and *Fomes fomentarius* and their functions as insect attractants. Journal of Chemical Ecology. 1999;**25**:567-590. DOI: 10.1023/A:1020958005023

[35] Siitonen J. Forest management, coarse woody debris and saproxylic organisms: Fennoscandian boreal forests as an example. Ecological Bulletins. 2001;**49**:11-41

[36] Artsdatabanken. Norsk rødliste for arter 2021. 2021. Available from: https://www.artsdatabanken.no/lister/rodlisteforarter/2021/

[37] Storaunet KO, Rolstad J. Naturskog I Norge. En arealberegning basert på bestandsalder i Landsskogtakseringens takstomdrev fra 1990 til 2016. NIBIO rapport. 2020;**6**(44):37. (In Norwegian)

[38] Larsson A. (red) Tillståndet i skogen – rödlistade arter i ett nordiskt perspektiv. ArtDatabanken Rapporterar 9. ArtDatabanken SLU, Uppsala. 2011

[39] Stokland J, Eriksen R, Granhus A. Tilstand og utvikling i skog 2002-2017 for noen utvalgte miljøegenskaper. NIBIO rapport. 2020;**6**(133):69. (In Norwegian)

[40] Jonsson BG, Kruys N, editors. Ecology of woody debris in boreal forests. Ecological Bulletins 49. 2001. 283 p

[41] Blindheim T, Thylén A, Reiso S. Sviktende kunnskapsgrunnlag i skog. BioFokus-Rapport. 2019;**11**:33

[42] The Norwegian Society for the Conservation of Nature. Sammenhengende Villmark (Continuous Wilderness). Ed. by Siri Tollefsen. 2020. 12 p. e-book ISBN 978-82-90895-90-2, printed ISBN 978-82-90895-89-6

[43] Junninen K, Kouki J. Are woodland key habitats in Finland hotspots for polypores (Basidiomycota)? Scandinavian Journal of Forest Research. 2006;**21**(1):32-40. DOI: 10.1080/02827580500530009

[44] Sverdrup-Thygeson A, Ims RA. The effect of forest clearcutting in Norway on the community of saproxylic beetles on aspen. Biological Conservation. 2002;**106**(3):347-357

[45] Koivula M, Vanha-Majamaa I. Experimental evidence on biodiversity impacts of variable retention forestry, prescribed burning, and deadwood manipulation in Fennoscandia. Ecological Processes. 2020;**9**:11. DOI: 10.1186/s13717-019-0209-1

[46] Wikars L-O. Effects of forest fire and the ecology of fire adapted insects [PhD thesis]. Sweden: University of Uppsala; 1997

[47] Wikars L-O. Dependence on fire in wood-living insects: An experiment with burned and unburned spruce and birch logs. Journal of Insect Conservation. 2002;**6**(1):1-12

[48] Hjältén J et al. Saproxylic insects and fire. In: Ulyshen M, editor. Saproxylic Insects. Zoological Monographs. Vol. 1. Cham: Springer; 2018. pp. 669-691

[49] Naturvårdsverket. Åtgärdsprogram för bevarande av brandinsekter i boreal skog. (Action plan for favouring red-listed fire-dependent insects in boreal forest in Sweden). Report 5610, 77 p. (in Swedish with English summary). 2006

[50] König M, Sallmén N, Westling U. Redovisning av ätgärdsprogram för bevarande av brandinsekter i boreal skog, 2016-2019. (conclusions from action plan for favouring red-listed fire-dependent insects in boreal forest in Sweden). Naturvårdsverket, Report. 2024. 23 p. (in Swedish)

[51] Finnish Forest Association. 2016. Available from: https://smy.fi/en/management-by-fire-increases-forest-biodiversity/

[52] Finnish Forest Association. 2018. Available from: https://smy.fi/en/finland-has-a-problem-with-too-few-forest-fires-to-promote-biodiversity-burned-down-areas-should-be-protected/

[53] Forest.fi. Newsletter 15.12.2023. Available from: https://forest.fi/article/wildfires-in-russia-affect-nature-on-finlands-eastern-border/

[54] Bakke A. Virkninger av skogbranner på billefaunanen. Rapport fra Skogforsk. 1996;**3**(96):1-20

[55] Brandrud TE, Bratli H, Sverdrup-Thygeson A. Dokumentasjon av sopp, lav og insekter etter Frolandsbrannen. Oppdragsrapport fra Skog og landskap 06/2010. Norsk institutt for skog og landskap. 2010

[56] Jansson U, Olberg S. Brannfelt—utkast til faktaark for kartlegging av naturtyper. Oslo: BioFokus; 2014

[57] Olberg S, Reiso S. Kartlegging av insekter på eldre brannflate ved Trynåsen, Bolkesjø. In: Biofokus-Rapport 2022-019. Stiftelsen: Biofokus. Oslo; 2022

[58] Rolstad J, Blanck Y, l. og Storaunet, K. O. Fire history in a western Fennoscandian boreal forest as influenced by human land use and climate. Ecological Monographs. 2017;**87**(2):219-245. DOI: 10.1002/ecm.1244

[59] Mico E. Saproxylic insects in tree hollows. In: Ulyshen M, editor. Saproxylic Insects. Zoological Monographs. Vol. 1. Cham: Springer; 2018. pp. 693-727

[60] Pilskog HE et al. Long-lasting effects of logging on beetles in hollow oaks. Ecology and Evolution. 2018;**8**(20):10126-10137. DOI: 10.1002/ece3.4486

[61] Skarpaas O, Diserud O, Sverdrup-Thygeson A, Ødegaard F. Predicting hotspots for red listed species: Multivariate regression models for oak associated beetles. Insect Conservation and Diversity. 2011;**4**(1):53-59. DOI: 10.1111/j.1752-4598.2010.00109.x

[62] Sverdrup-Thygeson A. Oaks in Norway: Hotspots for redlisted beetles (Coleoptera). In: Proceeding of the 5th Saproxylic Beetle Conference, 14-16 June 2008. Lüneburg, Germany: Pensoft Publishing; 2009

[63] Direktoratet for naturforvaltning. Handlingsplan for utvalgt naturtype hule eiker. 2012. 80 p. (In Norwegian with English Abstract and Summary)

[64] Hatlevoll K et al. Nasjonal overvåking av hule eiker: resultat av andre omløp. MINA fagrapport 62. Norges miljø- og biovitenskapelige universitet, Fakultet for miljøvitenskap og naturforvaltning. (In Norwegian with English summary, 36 p). 2019

[65] Hatlevoll K. Beetles in hollow oaks: the effects of traits on community structure [Master's Thesis]. Norwegian University of Life Sciences. 2020. p. 29

[66] Sverdrup-Thygeson A, Skarpaas O, Odegaard F. Hollow oaks and beetle conservation: The significance of the surroundings. Biodiversity and Conservation. 2010;**19**:837-852

[67] Wetherbee R, Birkemoe T, Skarpaas O, Sverdrup-Thygeson A. Hollow oaks and beetle functional diversity: Significance of surroundings extends beyond taxonomy. Ecology and Evolution. 2020;**10**(2):819-831

[68] Ranius T, Hedin J. The dispersal rate of a beetle, *Osmoderma eremita*, living in tree hollows. Oecologia. 2001;**126**:363-370

[69] Hilszczanski J, Jaworski T. Surrogate tree cavities: Boxes with artificial substrate can serve as temporary habitat for *Osmoderma barnabita* (Motsch.) (Coleoptera, Cetoniinae). Journal of Insect Conservation. 2014;**18**:855-861

[70] Hylander K, Ehrlén. The mechanisms causing extinction debts. Trends in Ecology & Evolution. 2013;**28**(6):341-346. DOI: 10.1016/j.tree.2013.01.010

[71] Jönsson AM, Harding S, Krokene P, Lange H, Lindelöw Å, Økland B, et al. Modelling the potential impact of global warming on *Ips typographus* voltinism and reproductive diapause. Climatic Change. 2011;**109**:695-718

[72] Hannerz M, Ekberg I, Norell L. Variation in chilling requirements for completing bud rest between provenances of Norway spruce. Silvae Genetica. 2003;**52**(3-4):161-168

[73] Venäläinen A, Lehtonen I, Laapas M, Ruosteenoja K, Tikkanen OP, Viiri H, et al. Climate change induces multiple risks to boreal forests and forestry in Finland: A literature review. Global Change Biology. 2020;**26**(8):4178-4196. DOI: 10.1111/gcb.15183. Epub 2020 Jun 13

[74] Della Rocca F, Bogliani G, Breiner FT, et al. Identifying hotspots for rare species under climate change scenarios: Improving saproxylic beetle conservation in Italy. Biodiversity and Conservation. 2019;**28**:433-449. DOI: 10.1007/s10531-018-1670-3

[75] Mirea MD, Miu IV, Popescu VD, Brodie BS, Chiriac S, Rozylowicz L. Priority conservation areas for protected saproxylic beetles in Romania under current and future climate scenarios. Biodiversity and Conservation. 2024;**33**:2949-2973